Sense and Nonsense of Statistical Inference

POPULAR STATISTICS

a series edited by

D. B. Owen

Department of Statistics
Southern Methodist University
Dallas, Texas

Nancy R. Mann

Biomathematics Department
University of California at Los Angeles
Los Angeles, California

Sense and Nonsense of Statistical Inference

Controversy, Misuse, and Subtlety

Chamont Wang

Department of Mathematics and Statistics
Trenton State College
Trenton, New Jersey

Library of Congress Cataloging-in-Publication Data

Wang, Chamont.
    Sense and nonsense of statistical inference : controversy, misuse, and subtlety / Chamont Wang.
        p.   cm. -- (Popular statistics ; 6)
    Includes bibliographical references and index.
    ISBN 0-8247-8798-6 (alk. paper)
    1. Science--Statistical methods.   2. Science--Philosophy.
3. Mathematical statistics.   I. Title.   II. Series.
Q175.W276   1993
507.2--dc20
                                                                92-35539
                                                                CIP

Marcel Dekker, Inc.
270 Madison Avenue, New York, New York 10016

When in doubt, tell the truth.
—Mark Twain, 1835–1910

Preface

This book concerns the misuse of statistics in science. Before and during the preparation of the manuscript, we witnessed a growing interest among the public in the exposure of misused statistics in a variety of sources: news stories, advertisements, editorials, federal reports, and academic research. This book presents examples from these areas, but its primary focus is the abuse and misconception of statistical inference in scientific journals and statistical literature.

Currently there are about 40,000 scientific journals being published regularly (*The New York Times*, June 13, 1988). While it is often claimed that science is self-correcting, this belief is not quite true when applied to the use of statistical generalization, probabilistic reasoning, and causal inference in scientific investigation. The examples of misused statistics in this book (most of which were produced by prominent scientists or statisticians) show that a large number of scientists, including statisticians, are unaware of, or unwilling to challenge, the chaotic state of statistical practices.

The central theme of this book is that statistical inference, if used improperly, can do more harm than good to the course of science. Accordingly, the book is written for the following audiences:

1. Graduate students of probability and mathematical statistics who will eventually teach and apply textbook statistics to real-life data. Many

scientists and statisticians stumble in their application of seemingly innocuous statistical techniques. We hope that our students will not have to carry these mistakes to the next generation.

2. Teachers of statistics who would like to understand better the limitations of statistical inference and would like penetrating examples to enrich their instruction. The examples in this book and the related discussions may well change the way they teach.

3. Empirical scientists who use statistical techniques to explore new frontiers or to support their research findings. In certain scientific disciplines (such as economics, educational research, psychology, biomedical research, and social science), researchers have been using sophisticated statistical methods, encouraged in part by the reward systems in their fields. Such researchers may find the discussions herein to be relevant to their academic endeavors.

Two additional audiences for whom this book will be informative are:

1. Applied statisticians who enjoy reading intellectual debates and would like to broaden their views on the foundation of statistical enterprise.

2. Theoretical statisticians who are curious about how their inventions have been abused by applied statisticians, empirical scientists, and other theoretical statisticians.

In recent years there is growing impatience on the part of competent statisticians (like Professor David Freedman of the University of California at Berkeley) with misused statistics in scientific and statistical literature. Indeed, all empirical scientists and academic statisticians who plan to conduct their business as usual now run some risk of finding their publications under attack.

Since its early development in the late 19th and early 20th centuries, statistics has generated waves of interest among empirical scientists. As a new branch of science, statistics was intended to deal with the uncertainty in the data. But, in many instances, statistical methods may actually *create* new kinds of uncertainty in scientific investigation.

These incidents are, nevertheless, quite normal in any young discipline. This book is thus an attempt to examine the "growing pains" of the statistical profession. For this purpose, the book is organized around the abuse and misconception of the following modes of statistical reasoning:

1. Tests of significance
2. Statistical generalization
3. Statistical causality
4. Subjective inference

Chapter 1 is devoted to the ubiquitous statistical tests. Testing of statistical hypotheses, as it appears in the literature, is at the center of the teaching and practice of statistical reasoning. However, in the empirical sciences, such tests are often irrelevant, wrong-headed, or both. The examples in this section are chosen from the biomedical sciences, economics, the behavioral sciences, and the statistical literature itself.

Chapter 2 discusses conflicting views of randomization, which, according to orthodox statisticians, is the very signature of statistical inference. In practice, randomization has two functions: (i) generalization and (ii) causal inference. Both functions have close connections with induction, a notorious topic in epistemology (the philosophy of science). Part of Chapter 2 is thus devoted to certain aspects of induction and epistemology, and also relates to the discussion in Chapter 3, as well as to a variety of topics concerning subjective knowledge and objective experimentation (Chapters 5–8).

Chapters 3 and 4 give examples of fallacious practices in statistical causal inference. Special emphasis is placed on regression models and time-series analysis, which are often used as instant formulas to draw causal relationships. In addition, Chapter 4 complements a motto of which all users of statistics should be aware: "No Causation without Manipulation" (Holland, 1987, *JASA*).

Chapters 5–8 are devoted to favorable accounts of observational studies in soft sciences, and to human judgment in nonexperimental settings. Contrary to a popular belief among scientists, sometimes subjective knowledge is more reliable than "objective" experimentation. Collectively, these chapters advocate the development of a critical eye and an appreciative mind toward subjective knowledge.

•   •   •   •   •   •   •   •   •

For many years, I felt cheated and frustrated by the application of so-called statistical models and by the practice of applying hypothesis testing to nonexperimental data. I am now less hostile to those statistics, because if used skillfully they can be intellectually illuminating. This book presents, therefore, not only criticism but also appreciation of uses of statistical inference.

To emphasize the constructive use of statistics, I have tried to include some illustrative examples: a modern version of Fisher's puzzle; graphic presentation and time-series analysis of Old Faithful data; Shewhart control chart; descriptive statistics; Bayesian analysis; chi-square test; parameter design in quality control; random inputs to a feedback system; probabilistic algorithms for global optimization; nonlinear modeling; spectral estimation; and quantum probability. These examples are intended to illustrate that *statistics is infinitely rich in its subtlety and esoteric beauty*.

Nevertheless, I believe that the practice of statistical inference has drifted into never-never land for too long. Therefore, this book concentrates on the misuse rather than on the bright side of statistical inference. A theme running through most of the examples is that the scientific community (and our society as a whole) will be better off if we insist on quality statistics. Individuals who share the same belief are urged to bring similar examples to light.

A good way to start is to have a critical mind when reading this book. Over the years, I have learned that the best way to deal with one's mistakes is not to hide them, but to laugh at them. For this reason, the author has admitted some of his own blunders (in Chapters 1, 2, 3, 5, 7, 8, and Intermission) for visible review.

So, although this book is about statistics, it is also about people, because people are at the center of defending or unraveling the integrity of the statistical profession. With this in mind, we have tried to illustrate the human side of statistical endeavor (whenever there is a fair prospect of fun) by providing comments and examples from our own recollections and the recollections of others. With all of these efforts and explanations, we hope that you can loosen up and enjoy a journey that may surprise you with some unexpected twists.

Chamont Wang

Acknowledgments

The author is deeply indebted to Albert J. Byer for his careful reading of the early drafts of Chapters 1 and 2 and parts of Chapters 3 and 4. His criticisms have substantially improved the final form of these chapters. Helpful comments were also provided by Philip Beardsley, Yvonne Kemp, and Lee Harrod. Finally, the author is most grateful to Herbert Spirer of the University of Connecticut for many of his suggestions which led to substantial improvements in the presentation.

Contents

Sense and Nonsense of Statistical Inference

Chapter 1

Fads and Fallacies in Hypothesis Testing

I. EXAMPLES: THE t-TEST

In an official journal of the American Heart Association, *Circulation*, Glantz (1980) presented a statistic on the misuse of statistics:

> Approximately half the articles published in medical journals that use statistical methods use them incorrectly.

Glantz's shocking article prompted the editors of *Circulation* to revise their review mechanism to help assure proper use of statistical methods in papers they publish. But it remains unknown whether other medical journals have taken the same action.

Most of the errors tabulated in the above study involved inappropriate use of the Student t-test. It is ironic that on one hand research statisticians have been meticulously using advanced mathematics (such as measure theory, topology, functional analysis, stochastic processes) to consolidate the foundation of statistical science; while on the other hand scientific articles still contain massive misuses of the Student t-test.

A further irony is that not only general scientists misapply statistical tests, but so do professional statisticians. An example was the figure given for the amount of time American students spent on their homework. According to the *Trenton Times* (November 1984, Trenton, New Jersey; and Bureau of Census, 1984), the statisticians at the Census Bureau reported that:

For Blacks, the total was 5.6 hours per week, compared with 5.4 for Whites,

and concluded that:

the overall difference between White and Black students, while small (about 12 minutes) is statistically significant.

There are several pitfalls in this report: (1) Twelve minutes a week (about 2 minutes a day) is not significant in the ordinary sense. (2) The observed significance level is a function of sample size. The sample size in the survey was about 60,000. When the sample size is this big, even 12 seconds a week may appear highly statistically significant. (3) The measurement of hours of homework is based on students' self-report, which is not very reliable; even professors cannot estimate with good accuracy the hours of their course preparation. Worse than that, the students in the survey, according to the Bureau's report, were aged 3 to 34 years! Based on this group of students and on the poor quality of the measurement, it is hard to imagine why the statisticians at the Bureau had to tell the public that the difference (12 minutes per week) was "statistically significant."

In sum, the Bureau's report relates to a typical confusion that equates "statistical significance" with "practical significance," a confusion that plagues many branches of empirical sciences.

A case study is the burgeoning literature on a psychometric measurement introduced in 1974 by Sandra Bem, then a psychologist at Stanford University. Bem explored the concept of psychological androgyny for people who are both masculine and feminine [andro: king; gyny: queen], both assertive and yielding, both instrumental and expressive. Bem's paper opened a new field of social psychological research. An impressive sequence of papers was written on the subject Bem pioneered. Many of them were attempts to find associations between androgyny and a wide range of variables: psychological health, personal adjustment, self-esteem, achievement, autonomy, and so on.

Bem's original idea was fruitful, and on the surface her paper appeared to be a solid study backed by modern statistical tests. However, a close look at the paper reveals that it contains disastrous misuses of statistical tests, such as (1) using the t-test to classify a personality trait as masculine or feminine, and (2) using the t-test to quantify the concept of androgyny as the absolute value of the difference between the masculinity and femininity scores. [The t-test and a number of related models will be used to illustrate a flat-Earth theory and stone-age measurements in a psychometric study. See Chapter 5, Section II.]

As a matter of fact, Bem used "statistical significance," instead of "practical significance" to establish a psychometric measurement and thereby ran into various problems. For example, both M and F scores contain respectively the items "masculine" and "feminine," which measure *biological* androgyny, not *psychological* androgyny.

I am somewhat sympathetic to Bem because most statistics books do not differentiate "practical significance" from "statistical significance." Bem is only one of many victims.[1]

II. A TWO-STAGE TEST-OF-SIGNIFICANCE

In 1975 Box and Tiao published a paper entitled "Intervention Analysis with Applications to Economic and Environmental Problems." The paper is widely considered as a milestone in the field of time-series analysis and has been frequently cited in literature. One application of the intervention analysis was to assess the impact of President Nixon's price controls. For this purpose, Box and Tiao (1975) built the following statistical model (where SE's are the standard errors related to the estimated coefficients in the model):

$$Y_t = -.0022 * X_{1t} - .0007 * X_{2t} + [(1-.85B)/(1-B)] * a_t$$

SE: .0010   .0009   .05

where Y_t = the consumer price index at time t; X_{1t} and X_{2t} correspond, respectively, to phase I and phase II of Nixon's price control, B is the backward operator, and a_t is the random noise. To build this model, as one can expect, we need intricate mathematical analysis and a relatively high-powered computer.

From the model, the authors reported, "The analysis suggests that a real drop in the rate of the CPI (Consumer Price Index) is associated with phase I, but the effect of phase II is less certain." To put it in statistical jargon, the effect of phase I control is significant, while the effect of phase II is not.

In 1983 I presented this example to my students in a time-series class. I was very proud that statistical tests were used to address important issues such as Nixon's price control. However, several graduate students majoring in economics commented that few (if any) in business schools would agree that either phase I or phase II of Nixon's price control is significant.

The students were courteous enough not to embarrass their instructor. But I felt stupid, and the event triggered a soul-searching for the nature of statistical tests. After years of reflection, I believe that statistical users will be better off if they take note of a two-stage test-of-significance as follows:

Step 1: Is the difference practically significant?

If the answer is NO, don't bother with the next step:

Step 2: Is the difference statistically significant?

This simple strategy certainly conflicts with the orthodox teaching: "Don't look at the data before you test!" (see Section V, DATA SNOOPING). Nevertheless it will serve statistics users in two ways: (1) Mistakes such as those made by the psychometricians and the Bureau of Census may be avoided. (2) Resources will not be wasted on building useless models which are not practically significant.

A question about this two-stage test-of-significance has been frequently asked: How do we know if a difference is "practically significant?"

The answer is that you have to use subject-matter knowledge. In general, there is no statistical formula that is readily applicable.

For instance, let X and Y be the price tags of something unknown, and assume that $Y - X = 15$ cents. Is this difference practically significant? It depends. If we are talking about the auto insurance premiums charged by two different companies, then the difference is not significant. On the other hand, if we are talking about the increment of the postage of an ordinary letter, then the difference is highly significant.

In certain cases, the appreciation of the "practical significance" can be subtle. Here is an example (Freedman, Pisani, and Purves, 1978):

> In the 1960s and 1970s, there has been a slow but steady decline in the Scholastic Aptitude Test scores (SAT). For the verbal test, the average in 1967 was about 465; by 1974, the average was down to about 445. The SD, however, has remained nearly constant at 100, and histograms for the scores follow the normal curve.

Now the question: Is the drop in SAT, $445 - 465 = -20$, practically significant?

On the surface, a 20-point drop may not seem like much, but it has a large effect on the tails of the distribution. For example, in 1967 and 1974 the percentages of students scoring over 600 was about 9% and 6%, respectively. The difference here is $9\% - 6\% = 3\%$, which still does not seem like much. But let's think about the percentage a little bit further. First, there are millions of students taking the SAT each year. Suppose that there are four million students. Three percent of them would translate into 120,000. That's a lot.

Second, students who score over 600 are more likely to be scientists, engineers, managers, intellectuals, or future political leaders. If we lose 3% of such students every 7 years ($1974 - 1967 = 7$), a few decades later this country will be in big trouble. Therefore, the drop in SAT scores, although only 20 points on the average, is practically highly significant.

We now move back to the case of Nixon's price control. In an econometric seminar at Princeton University, I met Professor Tiao and asked him specifically if Nixon's price control was significant in the ordinary sense. Professor Tiao looked stunned. It appeared that it was the first time somebody had asked him this question. His response was quick and right on the mark, though: he is only a statistician, so I should ask the question of people in business schools.

In Fig. 1 we include a graphic display of the consumer price index from January 1964 to December 1972.

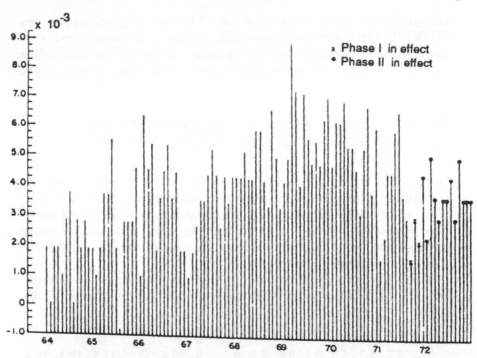

Figure 1 Consumer price index from January 1964–December 1972. Copyright © 1987 by the American Statistical Association. Reprinted by permission.

Note that at the right end of the graph, the black dots (phase II control) are running at about the same level as those before the phase I control. Therefore, phase II control did not appear to have reduced the inflation rate. Phase I control, on the other hand, appears effective. But a question is: If phase I was indeed effective, why was phase II needed? Further, if either control was effective, why was neither used by later administrations?

A motivation behind the proposed two-stage test-of-significance was that statistical methods usually take the place of a scientist's instinct and subject-matter knowledge. The situation often gets worse if complicated statistical models are used in empirical studies, where subject-matter knowledge is relatively weak. An example is a statistical model called discriminant function. The function is popular among behavioral scientists (e.g., Bem, 1979). This model depends on complex multivariate analysis formulas that general scientists are likely to have trouble understanding. Yet some researchers display a fondness for the complicated model and the terms like discriminant power, two-stage discriminant analysis, etc. (See an in-depth discussion of Bem's study in Chapter 5, Section II.) But a simple fact is that the statistical discriminant function is a generalized version of the t-test—it concerns only "statistical significance," not "practical significance." When dealing with the problems of classification, scientists should fully utilize their instinct and their knowledge of the subject matter. Playing around with the discriminant analysis appears to be a research fashion, but the classification based on such analysis may well identify a biological trait as a psychological characteristic again.

III. MORE EXAMPLES: A KOLMOGOROV-SMIRNOV TEST AND SOME DIAGNOSTIC STATISTICS FOR MODEL BUILDING

In the tests of statistical hypotheses, an issue about P-value is very confusing. For instance, Freedman, Pisani, and Purves (to be abbreviated as FPP, 1978, A-19) wrote: "Moderate values of P can be taken as evidence for the null."[2] This practice is common in regression and time-series analysis where moderate P-values of diagnostic statistics are used as indicators for the adequacy of model fitting.

To have a closer look at the issue of "moderate values of P," we first consider the following example of the Kolmogorov- Smirnov test (Conover, 1980, p. 371):

$$X = 7.6, 8.4, 8.6, 8.7, 9.3, 9.9, 10.1, 10.6, 11.2$$

$$Y = 5.2, 5.7, 5.9, 6.5, 6.8, 8.2, 9.1, 9.8, 10.8, 11.3, 11.5, 12.3, 12.5, 13.4, 14.6$$

The null hypothesis is that *the two populations have identical distribution functions*. The test statistic for the two-tail test is .40. Based on this test statistic, Conover concluded: "Therefore H_o is accepted at the .05 level." He further worked out the P-value to be slightly larger than .20, which is moderate, and thus can be taken, according to FPP, "as evidence for the null." However, a graphic display of the data looks like that in Fig. 2.

Eyeball examination indicates that the two distributions have nearly the same mean but quite different variances (Gideon, 1982). Shapiro-Wilk *normality tests* yield P-values of 92% and 43% for X and Y distributions, respectively. Consider these P-values as evidence that the underlying distributions are normal. Fisher's F-test therefore applies, and the end result ($P = .0097$) rejects the null hypothesis that the two distributions have the same variances.

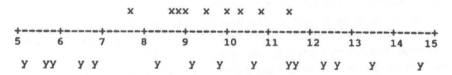

Figure 2   Graphical display of the data used in the Kolmogorov-Smirnov test.

After these analyses, amazingly, one still may argue that Conover's conclusion that "H_0 is accepted at .05 level" is not really wrong. The reasons are as follows: (1) The conclusion means that there are not enough data to reject the null hypothesis. (2) The fact that Fisher's F-test rejected the null hypothesis doesn't make the previous conclusion wrong. This is because the F-test is a more powerful test (if the normal-curve assumptions are correct). (3) The normality assumptions may not be valid, even if the Shapiro-Wilk tests yield P-values of 93% and 43%, respectively, for X and Y distributions.

Such arguments are bewildering to general scientists. But unfortunately all these arguments are correct. The following explanations may help. First, Freedman's statement that "moderate values of P can be taken as evidence for the null" is not quite true. In addition to a moderate value of P, one needs at least two other conditions in order to *accept* a null hypothesis: (1) there is substantial knowledge about the subject matter, and (2) the sample size is relatively large. To illustrate, consider the following example (FPP, 1978, p. 485):

The international Rice Research Institute in the Philippines is developing new lines of rice which combine high yields with resistance to disease and insects. The technique involves crossing different lines to get a new line which has the most advantageous combination of genes; detailed genetic modeling is required. One project involved breeding new lines for resistance to an insect called "brown plant hopper;" 374 lines were raised, with the results shown below.

	Number of lines
All plants resistant	97
All plants susceptible	93
Mixed	184

According to the IRRI model, the lines are independent; each line has a 25% chance to be resistant, a 50% chance to be mixed, and a 25% chance to be susceptible. Are the facts consistent with this model?

The answer according to FPP is that "chi-square = .3 with 2 degrees of freedom, (P $\cong$ 90%,) a good fit."

In our opinion, this conclusion was correct, but the reasons for the "good fit" offered by FPP were rather inadequate. For instance, if the numbers of lines were 0, 2, 2 (instead of 97, 93, and 184), then the P-value of the chi-square test would have been about 40%, and a "good fit" would have been declared by the same logic. In the case of the original conclusion, hidden behind the chi-square test is the feeling and the consensus among all scientists involved in the project. In practice, the scientists would look at the expected and the observed values:

Category	E = Expected Value	O = Observed Value	O − E
Resistant	93.5	97	+3.5
Susceptible	187	184	−3.0
Mixed	93.5	93	−0.5

By the comparison of the differences (O − E)'s in the last column, any reasonable person would agree that the genetic model underlying the calculation of the expected values is not far off from the reality. Another way to look at the data is by the calculations of the percentages:

E	O	O − E	(O − E)/E
93.5	97	+3.5	+3.7%
187	184	−3.0	−1.6%
93.5	93	−0.5	−0.5%
			Sum = 1.6%

The last column of the table shows that the differences between the observed and the expected are relatively small, especially for the bottom two categories. The plus and minus signs (albeit for only 3 values) also *indicate* that the differences may well be due to chance variation inherent in the genetic model. This chance variation can be formally quantified by the P-value of the chi-square test. In this case, P = 90%, therefore the null hypothesis that the model provides a good fit remains unchallenged.

A side note is that if nobody is bothered by the 3.7% deviation in the first category, then the chi-square test is *not* needed.

We now go back to Conover's non-parametric test which concluded that "H_0 is accepted at the .05 level." The statement certainly has nothing to do with "practical significance." But even on the issue of "statistical significance," the null hypothesis has been accepted on a very shaky ground. For instance, in Conover's book (p. 239) one can find a squared-ranks test for variances. The test is non-parametric, in the sense that it does not need the normal-curve assumptions. This test, as one can expect from the graphic display of the data, yields a P-value of .003. The null hypothesis is thus rejected without any reasonable doubt.

To sum up, Conover's conclusion that "H_0 is accepted at level" is not really wrong, *if* one accepts the current statistical terminology. This terminology is a typical case of "caveat emptor" (FPP, 1978, p. 350), i.e., the responsibility of misinterpretation is on the consumer.

•   •   •   •   •   •   •   •   •   •   •

In the November 1984 issue of *Science* magazine, the chi-square test was rated as one of the 20 discoveries (from 1900 to 1984) that changed our lives (Hacking, 1984). The article depicts the chi-square test as a formula to measure *the fit between theory and reality*. As we have seen in the previous rice example, this is too good to be true. To test the goodness-of-fit of a theory against experimental data, a scientist has to rely in large measure on his subject-matter knowledge, not only on a chi-square test (which concerns only statistical, not practical, significance).

Another misleading device of goodness-of-fit is the so- called "diagnostic checking" in regression or time-series modeling. Many statistical packages offer stepwise regression or automatic time-series modeling. The procedures are based on the search for the best choice of an array of diagnostic statistics. In this kind of modeling, as hinted by promoters of the software packages, all you have to do is enter the data and then the computer will take care of the rest.

To the insiders, such practices are in fact using a shot-gun approach in search of an essentially black box model which describes a very loose relationship between the inputs and the outputs but sheds little light on the true mechanism that generated the data. Statistical models of this sort, according to Geisser (1986, *Statistical Science*), "represent a lower order of scientific inquiry." Worse than that, such models may indeed stand on water.

To illustrate, let's cite a remarkable experiment from Freedman (1983). The experiment mimics a common practice in regression analysis where the underlying mechanism is relatively unknown (see references in Freedman's article). More precisely, a matrix was created with 100 rows (data points) and 51 columns (variables) so that all the entries in the matrix were pure noise (in this case, all entries are independent, identically distributed, and normal). The last column of the matrix was taken as the dependent variable Y in a regression equation, and the rest as independent variables. Screening and refitting in two successive multiple regressions yield the following results:

$R^2 = .36; P = 5 \times 10^{-4}$

14 coefficients were significant at the 25% level

6 coefficients were significant at the 5% level

Since the R-square and many coefficients in the regression equation are significant, the end product is therefore a law-like relationship between noise and noise.

The fact that such diagnostic tests may lead to a phantom relationship in regression or time-series analysis is a profound issue that will be discussed in more depth in Chapter 4 (and the related parts in Chapters 2, 3, and 6).

At this moment, we intend to point out only one fact. When Fisher and Neyman developed their theories for statistical testing, the test procedures required clearly-defined chance models (derived from substantial knowledge of subject matter). This logical sequence—model first, tests next—is reversed in "modern" diagnostic tests: the mechanism that generated the data is poorly understood, therefore the investigators use statistical tests to search *and* justify a model that might have described a relationship which never really existed.

IV. MECHANICAL APPLICATION OF STATISTICAL TESTS

In an article entitled "Nail Finders, Edifice, and Oz," Leo Breiman (1985) at the University of California, Berkeley, put forth a disturbing statement: "Many, if not most, statisticians in government and industry are poorly trained with a narrow and inapplicable methodology that produces limited vision, wizard-of- ozism, and edifice building." To the insiders of statistical practice, Breiman's assertion is unfortunately correct.

Lousy statistics can be found almost everywhere. Many of them arise, in our opinion, from mechanical use of statistical tests. The book *Statistics* by Freedman, Pisani and Purves (1974, 1975, 1978, 1980) launched a great effort to counteract such use. In the preface, they wrote,

> Why does this book include so many exercises that cannot be solved by plugging into a formula? The reason is that few real-life statistics problems can be solved that way.

With this in mind, we are compelled to include here a conversation between DeGroot and Lehmann in *Statistical Science*. DeGroot (1986) said, "Take a simple example of a chi-square test. Why is it universally used? Well, it is a simple procedure. You calculate it, and you don't have to think about it." DeGroot might have been joking around with the mechanical ".05" chance probability level. Nevertheless, most applied statistics books that we have seen exhibit essentially the same attitude (the list of the authors includes prominent statisticians at Chicago, Stanford, Harvard, Cornell, etc.).

The textbooks, as correctly pointed out by Kuhn (*The Structure of Scientific Revolution*, 1970, p. 144), are products of the aftermath of a scientific revolution, and they establish "the bases for a new tradition of normal science."

In the statistical discipline, this new tradition (as reflected in many popular textbooks) is the mechanical and mindless application of statistical methods. The situation has been greatly improved by the philosophy (and some techniques) of Tukey's EDA. But much remains to be done.

As an example of the chi-square test, a speaker at an econometrics seminar at Princeton said that he used the chi-square test with sample size near half a million (500,000) in a cross-sectional study for a big bank. It was interesting to see that this scholar, who is so careful in every detail of the proof by Lebesgue integration and functional analysis, is the same scholar who used a "simple" chi-square test in such a fuzzy manner. It was equally amazing that nobody except this author tried to point out that it was not right to do a chi-square test this way. For example, if we increase the sample size 100 times, the numerator in the chi-square test will increase about 10,000 times while the denominator increases only 100 times. This leads to the rejection of any reasonable null hypothesis.

In contrast to using the chi-square test without thinking about it, the following statement by Lehmann is a wise suggestion for any statistics users (DeGroot, 1986):

> [Many] Statistical problems don't have unique answers. There are lots of different ways of formulating problems and analyzing them, and there are different aspects that one can emphasize.

This statement reveals an important aspect of scientific activities: the intuitive and unpredictable way scientists actually work. In a sense, the term "scientific method" is misleading. It may suggest that there is a precisely formulated set of procedures that, if followed, will lead automatically to scientific discoveries.[3,4] There is no such "scientific method" in that sense at all.

V. DATA SNOOPING

Many statistics textbooks we have come across are loaded with formulas and statistical procedures about the so-called one-tail tests, two-tail tests, right-tail tests, and left-tail tests. Statistics users who are more conscientious about the validity of their results thus often ask this question: Which test should I use? right-tail or left-tail? one-tail or two- tail?

The answer to this question is, surprisingly, not contained in most books. It is like a professional secret prohibited to non-statisticians or to entry-level statistics majors. The correct answer, I finally found out, was: It is up to the users. This answer does not seem to provide much guidance. Further, it is even more confusing that, according to the orthodox teaching of the Neyman-Pearson school, one cannot look at the data before one sets up a one-tail or two-tail test.

The orthodox teaching of "Don't look at the data before you test" has something to do with the probabilities of the type I and type II errors in hypothesis testing. The probabilities are meaningful, according to the school of Neyman and Pearson, only before the data are examined. This may sound strange, but consider this example. A box contains 30 red and 70 blue marbles. A marble is picked at random from the box. Before we look at the marble in hand, the probability that it is blue is 70%. But once we looked at it, the probability is either 0% (a red marble) or 100% (a blue marble), which means probability theory does not apply anymore.

Note that the orthodox teaching casts doubt on the legitimacy of the two-stage test-of-significance and the calculations in many published results. Therefore it is worthwhile to examine how statisticians in other schools think about the issue.

Dempster (1983) mentioned that "taken seriously, Neyman's theory suggests that all the statistician's judgment and intelligence is exercised before the data arrives, and afterwards it only remains to compute and report the result." Since Dempster views reasoning in the presence of data to be the crux of data analysis, he cannot "take the theory seriously either as an actual or normative description of the practice of Confirmatory Data Analysis."

Huber (1985) adopted the orthodox teaching as a tool complementary to visual analysis and mentioned that he always seems to get P-values in the range .01 to .1 if the data are examined before the test. Huber's explanation of this phenomenon is: if P-value is small (P < .01), then he seems to accept uncritically the feature as real without bothering to test; on the other hand, if P-value > .1, he would not see the feature and therefore would not test.

Freedman et al. (1978, p. 494) said, "Investigators often decide which hypothesis to test, and how many tests to run, only after they have seen the data." To Freedman, this is a "perfectly reasonable thing to do," but "the investigators ought to test their conclusions on an independent batch of data."

On a related issue of whether to use one-tail or two-tail tests, FPP wrote: "In fact, it doesn't matter very much whether an investigator makes a one-tailed or a two-tailed z-test, as long as he tells you which it was. For instance, if it was one- tailed, and you think it should have been two-tailed, just double the P-value."

Freedman's argument is intellectually persuasive and would certainly benefit scientists in their data analyses. However, many Bayesians may raise a wall against such practices (see, e.g., *JASA*, Vol. 82, pp. 106–139; also Berger and Delampady, 1987). The issue at stake is very nasty, and scientists are advised to stay away from this mess. After all, doubling the P-value is implicitly in good harmony with the Neyman-Pearson paradigm (see more explanations near the end of Section VIII). Statistics users need not panic at the objections of radical Bayesians.

VI. AN APPRECIATION OF "NON-SIGNIFICANT" RESULTS

The original intention of statistical testing was to protect scientists from drawing false conclusions in the face of chance effect. As apparent from the examples shown in this chapter, this cure often is worse than the disease.

Bewildered by statistical tests, scientists often underestimate the potential of "non-significant" results. This tendency is another disservice of our profession to the course of science. Some examples are as follows.

In a survey of articles published in leading medical journals, Pocock et al. (1987, p. 431) tabulated 221 results, of which 91 were reported significant and 130 non-significant. Pocock observed that abstracts of research publications are very important in transmitting new knowledge to scientific community. But intimidated by the so-called "non-significant" statistical tests, scientists often do not report in the abstract the findings that are associated with large P-values.

To see the extent of this consequence, Pocock noted that only 25% of the 130 non-significant comparisons were included in the abstracts of the studies (whereas 70% of the 91 significant comparisons were included). It is possible that the remaining 75% of the results with large P-values are practically significant, but were overlooked simply because they are not statistically significant.

Pocock also observed that out of 45 clinical trials surveyed, only six (13%) made use of confidence intervals. A conclusion was drawn by Pocock:

> Because of the obsession with significance testing in the medical literature, authors often give insufficient attention to estimating the magnitude of treatment differences.

This phenomenon is not unique in medical literature; the problem is also prevailing in other empirical studies.

In a meta-analysis of psychological individualism, Archer and Waterman (1988) surveyed 88 comparison tests, of which 65 (74%) reported no significant differences. In another survey of 574 tests of association, Archer and Waterman counted 247 non-significant and 327 significant outcomes. In both studies, the actual magnitudes of differences and correlations were *not* indicated.

We suspect that many so-called "non-significant" results may in fact contain important information if the investigators had looked at the magnitudes of the differences. Lest statistical jargon lead researchers astray, it should be made explicit that "statistical insignificance" simply means that noise overwhelms the signal and that one needs a larger sample or better measurements. In many cases, increasing sample size will eventually lead to a "statistically significant" result. In other cases, the noise/signal ratio is so large that the investigators have to go back to the drawing boards.

We now summarize the discussions of significance tests in the tree diagram shown in Fig. 3 (this is an extension of the two-stage test-of-significance). In Fig. 3, scientists are advised not to give up the research in case the observed difference is not significant but a strong theory says otherwise. The reason is that many things can go wrong in the stages of data collection and data analysis. An example follows.

Modern astrophysicists conclude that our universe is filled with ultra-smooth CBR (cosmic blackbody radiation). They also believe that CBR is the remnant of thermal radiation from a condensed primordial universe—the hot

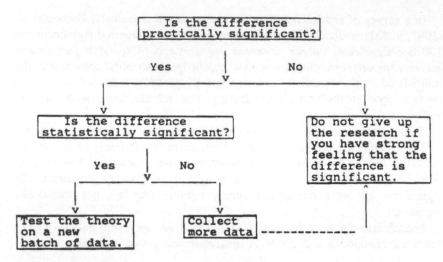

Figure 3   Strategy for a multistage test of significance.

Big Bang. Theoretical physicists predict that primordial fluctuations in the mass density, which later became the clusters of galaxies that we see today, should have left bumpiness on the blackbody radiation. The prediction is of great importance to the study of the Big Bang model and has thereby generated numerous experiments and articles on the subject. (See references in Wilkinson, 1986.)

In order to test the theory, CBR has been measured with sophisticated radiometers on the ground, in balloons, and in satellites. A statistical model of the measurement can be described as follows:

$X_1, X_2, \ldots, X_n$ are i. i. d. (independent and identically distributed) normal with mean 0 and variance $f+e$, where f is the strength of fluctuation and e corresponds to the background noise.

One-sided 95% confidence intervals (each equivalent to a family of hypothesis testings) have been calculated to measure the amplitude (f). But none of the estimated amplitudes has been proven to be statistically significant. The situation is rather disappointing because the best available technology has been fully explored. A question then arises: "What can scientists do now? Abandon the Big Bang theory?" No, not yet. The only thing which has been established is that blackbody radiation is extremely smooth. No more, no less. As a matter of fact, physicists dedicate themselves to a continual search for the fluctuation and a deeper understanding of the blackbody radiation. Their activities include better calibration of the instruments, the reduction of the background noise, and applications of sophisticated statistical methods, etc.

In one instance, some physicists attempted to capitalize on the findings in the existing studies. They proposed intricate arguments (a combination of Bayes Theorem, likelihood ratio, Legendre polynomials, type I and type II errors, etc.) to set upper limits for the amplitude of the model. Their arguments are very intelligent but were dismissed by this author on the ground that the prior distribution is sloppy and indeed, as they admitted, contradictory to existing knowledge. After a series of failures, these physicists still have no intention of giving up; they vowed that they will try other priors. They also are thinking of going to the South Pole to get better measurement of blackbody radiation.

Physics has long been admired by many as the champion of "exact science." In the field of astrophysics, however, this branch of the "exact science" is not all that exact. Nevertheless, their relentless pursuit of good theory and good measurement is highly respectable and is in sharp contrast to many empirical studies that rely mainly on statistical tests of significance.

VII. TYPE I AND TYPE II ERRORS:
FOR DECISION-MAKING

In a personal interview, Erich Lehmann (a la DeGroot, 1986, *Statistical Science*) quipped about the mechanical application of the .05 significance level as "obviously a silly thing to do." To avoid this silly practice, Lehmann emphasized the importance of the power function and "a better balance between the two kinds of errors," which are better known as the type I and type II errors in hypothesis testing.

It is hardly an overstatement to say that the type I and type II errors are dominant components of hypothesis testing in statistical literature. However, scientists usually perform statistical tests based only on type I error and seldom bother to check the type II error. This lack of adherence to the Neyman-Pearson paradigm can also be found in the following books, where the lack of adherence turned into the lack of respect.

Wonnacott and Wonnacott (1982) discussed some difficulties with the classical N-P (Neyman-Pearson) procedure and asked: "Why do we even bother to discuss it?" They concluded that the N-P theory is of little use in practice. In their book, FPP (1978) simply ignore the whole Neyman-Pearson theory. A question to Freedman is: How can scientists perform statistical tests without knowing type I and type II errors?

For this issue, consider the following example which I found successful in convincing some skeptics and would-be scientists of the importance of type I and type II errors. The issue is the safe usage level of a cheap but potentially harmful farm chemical. Assume the safe level is 10 units or less. So the null hypothesis is either (A) H_o: $\mu \leqslant 10$ or (B) H_o: $\mu \geqslant 10$. In case (A), the null hypothesis says that the chemical is at a safe level, while in (B) the null hypothesis says that the chemical is at a dangerous level. Users are often confused as to which null is more appropriate. Let's discuss this issue later. To set up decision rules, we need sample size n, the probability of type I error, and the estimated standard deviation (SD). Assume n = 25, SD = 4, and alpha = .01. For case (A), the rejection region is $\overline{X} > 11.994$. For case (B), the rejection region is $\overline{X} < 8.006$. Now we can make a decision whenever $\bar{x}$ is available. If $\bar{x} = 11.0$, then under (A) we conclude that H_0 remains unchallenged and that the chemical is safe; under (B) we also do not reject H_0 and conclude that the chemical is dangerous. If $\bar{x} = 8.5$, we reach the same conclusion for (A) or (B). *The set-ups of the right-tail or left-tail null hypothesis thus give us opposite conclusions!*

To clarify the confusion, let's compare the type I errors. For case (A), alpha is the error which farmers do not want to risk. For case (B), alpha is the error which consumers do not want to risk. Therefore, to protect consumers, we should choose the right-tail null hypothesis, even if the consumers have to pay more for the farm products. Another reason is: If we save money today, we may have to pay more to hospitals tomorrow. Hence (B) is the choice after we analyze the nature of the type I errors.

Some of my colleagues and students enjoyed the above discussion of type I errors and praised it as a "beautiful analysis." I then went further to discuss the alternative hypotheses and the type II errors.

Two years after I introduced the previous example in the classroom, I totally changed my mind. To highlight the point, let me quote a statement on the calculus of variations from a classic book by Bellman and Dreyfus (1962, p. 182):

> After the initial development of the calculus of variations by Euler, Lagrange, Weierstrass, Hilbert, Bolya, Bliss and others, the subject became, to a large extent, more fit for the textbook than the laboratory.

This is what I now feel about the type I and type II errors advocated by Lehmann among others. None of the analysis in the above example is necessary if we simply look at the 98% confidence interval ($\overline{X} \pm 1.992$). Whether $\bar{x} = 11.0$ or 8.5, we cannot conclude that the chemical is at a safe level. For decision-makers, a larger sample size and a confidence interval will certainly solve this problem without any help from the analysis of type I or type II errors.[5]

Decision-making based on the analysis of type I and type II errors does make sense in certain industrial quality control. But the practice is facing increasing challenge from some QC experts who put more emphasis on "zero defects" and the concept of "doing the things right at the beginning." (See more in Chapter 8, Section II.) This philosophy, in our observation, is gaining ground, and the whole industry of OC curves (including Military STD 414, 105D, etc.) may become obsolete in the near future.

In some ways, enthusiastic statistics users are convinced that the analysis of type I and type II errors are powerful tools for decision-making. For instance, if you look at the freshman books in business statistics (which are supposed to be relevant to real-life applications), you will often find several chapters on "decision theory" that deal with type I and type II errors. Students, on the other hand, regard these materials as part of "an unimportant, mechanical, uninteresting, difficult but required subject" (Minton, 1983).[6] We believe that these students are quite smart. Only those who lose their sanity completely would try to deal with type I and type II errors in any business setting.[7]

In short, it is a grave mistake to confuse statistical decision theory with ordinary decision theory, yet this is done all the time in business statistics books. In an ASA (American Statistical Association) panel discussion on the role of the textbook in shaping the business statistics curriculum, Hoyer (1987) decried that:

> The Neyman-Pearson decision theory is a logically indefensible and intellectually bankrupt decision mechanism.

We applaud this conclusion and hope that someday those who produce business statistics textbooks will take note and really ponder the true nature of the Neyman-Pearson theory of statistical hypothesis testing.

VIII. TYPE I AND TYPE II ERRORS: FOR GENERAL SCIENTISTS

The essence of the Neyman-Pearson theory is to provide criteria to evaluate the sharpness of competing tests and then opt for the tests that perform best (Lehmann, 1985). But this task is primarily a job belonging to theoretical statisticians, not practicing statisticians nor general scientists.

In fact, learned statisticians seldom compute type II error on a set of experimental data. They simply calculate P-values (and confidence intervals), and the conclusions are not less rigorous.

It is a common belief that type II error helps to determine the sample size in the design of experiments. However, the procedure is cumbersome and very artificial. An example can be found in a book called *Design of Experiments: A Realistic Approach* by Anderson and McLean (1974, pp. 4–6). In this case, the concern was the strength of two-different cement-stabilized base material. For this simple comparison, the investigators enlisted a special scheme, and, according to Anderson and McLean, "the sequential procedure recommended here or by Stein (1945) *must* be followed" [emphasis supplied]. This so-called sequential procedure, as one must be aware, has nothing to do with the actual magnitude of the difference. Rather, it concerns only the number of additional specimens to take. The procedure is complicated and involves contrived selection of beta-values (i.e., the probabilities of making the type II errors) that is hard to pin down and hardly makes intellectual sense. As a matter of fact, the width of a confidence interval can be used to determine the sample size, and it is definitely not true that the sequential procedure must be followed.

In more complicated designs (such as multiple comparisons or two-way layouts), the choices of sample size involve certain ingenious procedures developed by mathematical statisticians. Such procedures are of great esoteric interest to specialists but are seldom needed in data analysis. In our opinion, interval estimation (such as Tukey, Sheffé, or Bonferroni simultaneous confidence intervals) will not be less rigorous than the stopping rules based on type I and type II errors. In conjunction with the interval estimation, a "sequential procedure" should proceed as follows:

1. Examine the actual magnitudes of the differences in a multiple comparison.
2. Throw out the cases that are neither practically nor statistically significant.
3. Conduct a trimmed multiple comparison[8] which by this moment may be only a simple 2-population comparison.

This approach would apply to most statistical designs, such as

one-factor analysis
two-factor analysis
complete factorial experiments
incomplete factorial experiments
nested factorial designs
Latin square designs

Note that all these designs are special cases of regression models. The treatment effects thus can be assessed by the estimation of linear combinations of parameters.

In certain statistical tests (such as chi-square test, normality test, and runs tests) the use of the confidence interval is not appropriate; but do not expect that the type I and type II errors will help in the determination of the sample size. In most cases, a learned statistics user simply follows his feeling, and generally this works out better than many exotic designs which are based on the type II errors.

In our opinion, the accept-reject method is useful only for certain engineers in quality control, but irrelevant to the data analyses in general science. However, the accept-reject method and the calculation of the power function are repeatedly promoted in leading medical journals (see, e.g., Glantz, 1980; Der-Simonion et al, 1982; etc.). The reasons are given in Pocock et al. (1987, p. 430) and can be summarized as follows.

First, data snooping may introduce bias in the study; hence the intended size of a trial should be determined in advance. Second, it is *ethically* desirable to stop the study early in case clear evidence of a treatment difference has been found. Since repeated use of statistical testing increases the risks of a type I error, some stopping rules based on power calculations are recommended to researchers in biomedical science (see references in Pocock).

Pocock's reasons for power calculation in clinical trials are questionable for both ethical and technical reasons. For ethical reasons, most multiple comparisons involving human beings indeed can be (and should be) trimmed to two-population comparisons. Therefore interval estimation can be used to determine the sample size. Using human subjects in a complicated design only results in a waste of valuable resources.

For technical reasons, Pocock's recommendation is a two- stage procedure that requires testing to take place before the estimation of the contrast between parameter values. However, it is known (see, e.g., Brown, 1967; Olshen, 1973; and Casella and Hwang, 1991) that this two-stage procedure (testing first, estimation next) can destroy the probability statement of a confidence interval.

As shown in Olshen (1973), for a broad spectrum of experimental designs, the conditional probability of simultaneous coverage (given that the F-test has rejected the null hypothesis) is *less* than the unconditional probability of the coverage. Further, it is impossible to assign precisely a conditional probability of coverage, because that conditional probability depends on unknown parameters.

In conclusion, after much effort, we fail to see any convincing case in which the statistical scaffolding of power calculation is needed. At this moment, we believe that the power calculation is only a weak and often misleading procedure and that scientists may be better off to abandon the accept-reject method altogether.

In an ASA panel discussion at the Chicago annual meeting 1986, John Bailar (representing the *New England Journal of Medicine* and Bruce Dan (representing the *Journal of the American Medical Association*) endorsed the conventional .05 and type I, type II errors. Bailar and Dan also disliked the reporting of small P-values. Note that in scientific reporting, a statistical result is often marked with one star if $P < .05$, and two stars if $P < .01$. Bailor and Dan (1986) thus poked fun at the authors who reported small P-values of .0003 or .000026 or any number "followed by 3 or 4 stars."

Nevertheless, in the *Annals of Statistics*, Diaconis and Efron (1985, p. 905) reported a significance level of "2×10^{-5}, overwhelming proof against the hypothesis of fairness," and in *JASA*, Ethier (1982) reported "an approximate P-value of less than 2.3×10^{-6}, a strong indication of one or more favorable numbers."

Freedman et al. (1978, p. 494) discuss the historical origin of 5% and 1% in the chi-square table and conclude, "With modern calculators, this kind of table is almost obsolete. So are the 5% and 1% levels." Freedman et al. (p. 93) also reported "a minor miracle" for the appearance of an outlier five standard deviations away from the average.

In short, it appears that the "official" 5% and 1% significance levels are not held in high regard by some top-rated statisticians. Ordinary statistics users may now wonder about the following question: "Should a scientist report small P-values, to say 10^{-6}?"

The answer is a definite YES. For example, in an experiment of paired comparison (n = 100), an average of 6 grams was observed, and the difference was important to the subject-matter experts. However, the P-value is 4.9%, only slightly smaller than the 5% level. Assume that the scientist in charge is diligent and decides to run the experiment one more time but *under better control* (n = 100). The new average remains similar, yet the variability in the data is drastically reduced and the P-value is now only 10^{-6}. After all this work, the scientist would like to ask the gate-keepers of medical journals: "Can I put five stars to this result?"

Furthermore, assume that other scientists repeat the experiment under similar conditions. The average and SD of the reported P-values are respectively 10^{-6} and 10^{-7}. Now should the scientific community simply report that the result is significant at the 5% (or 1%) level?

In practice, the reporting of P-value is superior to the accept-reject method in the sense that it gives a useful measure of the credibility of H_0. This use of P-values has strong flavor of Bayesianism, but an exact interpretation of this measure is difficult and controversial. In a 1987 edition of *JASA* (Vol. 82, pp. 106–139; a whopping 34-page discussion), ten famous statisticians participated in a heated debate on the reconciliability of P-values and Bayesian evidence; and the conclusions were that a justification of P-value in terms of Bayesian evidence was found and lost. The debate of those ten statisticians is intriguing, and it appears that hardly any of those scholars came out of that mess completely clean.

For statistics users, the bad news is that a Bayesian calibration of P-value is very difficult (or nearly impossible), but the good news is that P-values are "reliable for the primary hypothesis in well-designed (for good power) experiments, surveys, and observational studies" (Morris, 1987, p. 132). For hypotheses of secondary interest, call a professional statistician.

Unlike the probabilities of type I and type II errors, P-value per se does not bear a frequentist interpretation of the calculated probability value. As a result, inference based on P-value alone is not grounded on probability theory, but rather on "inductive reasoning"—a very powerful yet often misleading scientific method (for more discussions of the merit and the problem of inductive reasoning, see Chapter 2, "Quasi-Inferential Statistics.")

In contrast to P-values, Neyman-Pearson's type I and type II errors concern a statistical procedure rather than a simple statistic; and inference based on N-P theory is deductive, not inductive. Berger and Delampady (1987) asserted that in certain testing procedures "formal use of P-values should be abandoned." One of their reasons is that P-values do not have a valid frequentist interpretation. This is true. But in practice, scientists usually compare P-values to 5% or 1% significance level. Inferences of this type then appear to be deductive, not inductive. In brief, P-value *enjoys* the power of inductive reasoning and at the same time establishes its legitimacy upon the strength of the theory of Neyman and Egon Pearson.

The above discussion appreciates an advantage of the N-P theory over Fisher's fiducial probability. But this advantage of the N-P theory disappears when scientists approach real-life data. Indeed, the tyranny of the N-P theory in many branches of empirical science is detrimental, not advantageous, to the course of science.

IX. CONCLUDING REMARKS

In practice a confidence interval corresponds to a *family* of hypothesis testings and thus is a more useful method. To certain theoretical statisticians, however, interval estimation is a rather simple procedure and has little intellectual appeal. Such statisticians therefore lavishly invest their academic freedom in the theory of hypothesis testing. Fortunately, this investment has generated many beautiful theories on the power properties of statistical tests (see, e.g., LeCam's Third Lemma in Hájek and Sidák, 1965).

In addition, these theories were extended to the Hájek-LeCam inequality and a new field of adaptive estimators that attain the lower bound of the inequality and thus provide optimal estimation. On other fronts, test statistics are used to construct new estimators. For instance, Wang (1986) derived, from test statistics, a minimum distance estimator for first order autoregressive processes.

For such reasons, the Neyman-Pearson theory remains an important topic in mathematical statistics. But regretfully, the N-P theory is only a theory. It is not a method, or at least not a method that cannot be replaced by better methods.

In certain branches of academic disciplines, according to Lehmann (a al DeGroot, 1986), "hypothesis testing is what they do most." We believe this is true, but it may not be a healthy phenomenon. In regard to the wide use of statistical tests, DeGroot (1986, p. 255) asked: *"Why is it, and should it be?"* In fact, Morrison, Henkel et al. (1970) and many veterans of hypothesis testing raised the controversy about the usefulness of statistical tests in behavioral science. They concluded:

Taken together, the questions present a strong case for vital reform in test use, if not for their total abandonment in research.

Henkel (1976), then an associate professor of sociology at the University of Maryland, further concluded that: "In the author's opinion, the tests are of little or no value in social science research."

In principle, we agree with Henkel and Morrison, although we believe that P-value can be used as an easy and powerful check for the difference between observed and expected values. Nevertheless, we deeply resent the way Neyman-Pearson theory is taught and applied. Most applied statistics books we find are loaded with "involuted procedures of hypothesis testing" (DeGroot, 1986, p. 255), as well as lengthy discussions of type II error which has little bearing on scientific inquiry. On the surface, those applied books look rigorous in their mathematics, but in fact *they are neither mathematical nor statistical, but only mechanical!* No wonder empirical scientists are often misled by statistical formulas and produce many results which earn little respect from serious scientists (see the examples of psychometric research in Chapter 1, Section I and Chapter 5, Section II).

As a comparison, literary criticism in history and good journalism, which are relentlessly grounded on hard evidence and insightful analysis, may bear more scientific spirit than many so-called "data analyses" carried out by psychometricians and perhaps by many other empirical scientists, too.

We believe, despite such questioning, the test of significance is in all probability here to stay (especially the nonparametric and goodness-of-fit tests), but its reappraisal for general scientists is also necessary. As long as leading statisticians keep on ignoring the true nature of statistical tests and keep teaching and writing introductory books for non-statistics students, the reappraisal is almost impossible. This is a solemn appeal to statisticians to look at the issue seriously, and this is the only way that the problem can receive the treatment it merits.

NOTES

1. Kruskal and Majors (1989) sampled 52 usages of "relative importance" from a diversity of disciplines such as physics, chemical engineering, biology, medicine, social sciences, and humanities. The sample was taken (by using a probability method) from a frame of 500-odd papers having *relative importance(s)* or *relative influence(s)* in the title. Out of the 52 usages, about 20% judged relative importance by observed significance levels (P-values).

    See also Problem #2 in a popular textbook (Hicks, 1982, p. 32). In this case, the difference between the observed value and the expected value is highly significant ($z = 3.24$, $P < .0005$), but practically inconsequential (($\overline{X} - \mu)/\mu = 2.9/18470 = .00015$).

2. Freedman et al. (1978) wrote: "P-value does not give the chance of the null hypothesis being right." It is odd that FPP are crystal-clear throughout Chapters 26–29 but give us a loose statement on A-19. I am somewhat relieved to know that Freedman, too, is a human being. The mistake, however, was corrected in Freedman, Pisani, Purves, and Adhikari (1991, 2nd edition, Norton).

3. "Systematic" application and evaluation of statistical methods are recommended. Mechanical application of statistics are misleading and intellectually offensive. See also Goldstein and Goldstein, 1978, p. 4.

4. A computer program, BACON (Langley, 1981), was used to rediscover Kepler's third law with a combination of Kepler's methods and sophisticated heuristics. BACON was also claimed to have rediscovered many other scientific laws. But the program, to my knowledge, has not yet discovered any useful law. We are willing to give BACON our best wishes, though.

5. Even the simple calculation of P-values (10% if $\bar{x} = 11$ and 4% if $\bar{x} = 8.5$) will make the discussions of the type I error in this example unnecessary.

6. Minton's discussion did not deal just with business statistics, but was more general in nature.

7. There are only two situations in which we believe the concept of type II errors is really useful: (1) comparison of sensitivities of control charts, and (2) acceptance sampling in quality control. However, the current trend in QC is a de-emphasis of acceptance sampling.

8. An ingenious way to conduct a trimmed multiple comparison is the classical FFE (fractional factorial experiment; see, e.g., Hicks, 1982). This is one of the greatest contributions of statistics quality control. The procedure uses formal inferential statistics. But FFE is different from the traditional methods such as Tukey's or Scheffé's procedures: FFE is more a technique for exploratory data analysis, and less a method for confirmatory analysis.

REFERENCES

Anderson, V. L. and McLean, R. A. (1974). *Design of Experiments, A Realistic Approach.* Marcel Dekker, New York.

Archer, S. L. and Waterman, A. S. (1988). Psychological Individualism: Gender Differences or Gender Neutrality? *Human Development*, Vol. 31, 65–81.

Bailar, J. and Dan, B. (1986). Interactions between Statisticians and Biomedical Journal Editors: A Panel Discussion at ASA Annual Meeting.

Bellman, R. and Dreyfus, S. (1962). *Applied Dynamic Programming.* Princeton University Press, Princeton, New Jersey.

Bem, S. L. (1974). The Measurement of Psychological Androgyny. *J. of Consulting and Clinical Psychology*, Vol. 42, 144–162.

Bem, S. L. (1979). Theory and Measurement of Androgyny: A Reply to the Pedhazur-Tetenbaum and Locksley-Colten Critiques. *J. of Personal and Social Psychology*, Vol. 37, 1047–1054.

Berger, J. and Delampady, M. (1987). Testing Precise Hypotheses. *Statistical Science*, Vol. 2, 317–335.

Box, G. E. P. and Tiao, G. C. (1975). Intervention Analysis with Applications to Economic and Environmental Problems, *JASA*, Vol. 70, 70–79.

Breiman, L. (1985). Nail Finders, Edifice, and Oz. *Proceedings of the Berkeley Conference in Honor of Jerzy Neyman and Jack Kiefer*, Vol. 1, 201–212, Wadsworth, Belmont, California.

Brown, L. (1967). The Conditional Level of the t Test. *Annals of Mathematical Statistics*, Vol. 38, 1068–1071.

Bureau of the Census (1984). Population Characteristics, *Current Population Reports*, Series P-20, No. 394.

Casella, G. and Hwang, J.T. (1991). Comment to "On the Problem of Interactions in the Analysis of Variance" by V. Fabian. *JASA*, Vol. 86, 372–373.

Conover, W.J. (1980). *Practical Nonparametric Statistics*. Wiley, New York.

DeGroot, M.H. (1986). A Conversation with Erich L. Lehmann. *Statistical Science*, Vol. 1, No. 2, 243–258.

Dempster, A.P. (1983). Purposes and Limitations of Data Analysis. *Scientific Inference, Data Analysis, and Robustness*, edited by G. E. P. Box, T. Leonard and C. F. Wu. 117–133. Academic Press, New York.

DerSimonian, R.,Charette, L. J., McPeek, B., and Mosteller, F. (1982). Reporting on Methods in Clinical Trials. *The New England Journal of Medicine*, Vol. 306, No. 22, 1332–1337.

Diaconis, P. and Efron, B. (1985). Testing for Independence in a Two-way Table: New Interpretations of the Chi-Square Statistic. *Annals of Statistics*, Vol. 13, No. 3 845–874.

Ethier, S. N. (1982). Testing for Favorable Numbers on a Roulette Wheel. *JASA*, 660–665.

Freedman, D. A. (1983). A Note on Screening Regression Equations. *American Statistician*, Vol. 37, 152–155.

Freedman, D. A., Pisani, R., and Purves, R. (1978). *Statistics*. Norton, New York.

Geisser, S. (1986). Opera Selecta Boxi. *Statistical Science*, Vol. 1, No. 1, 106–113.

Gideon, R. (1982). Personal Communication at the Department of Mathematical Sciences, University of Montana.

Glantz, S. A. (1980). Biostatistics: How to Detect, Correct and Prevent Errors in the Medical Literature. *Circulation*, Vol. 61, 1–7.

Goldstein, M. and Goldstein, I. F. (1978). *How We Know, an Exploration of the Scientific Process*. Plenum Press, New York.

Hájek, J. and Sidák, Z. (1967). *Theory of Rank Tests*. Academic Press, New York.

Hacking, I. (1984). Trial by Number. *Science*.

Henkel, R. E. (1976). *Tests of Significance*. Sage, Beverly Hills, California.

Hicks, C. R. (1982). *Fundamental Concepts in the Design of Experiments*, 3rd ed. Holt, Rinehart, and Winston, New York.

Hoyer, R. (1987). The Role of the Textbook in Shaping the Business Statistics Curriculum. A preprint.

Huber, P. (1985). Data Analysis: In search of an Identity. *Proceedings of the Berkeley Conference in Honor of Jerzy Neyman and Jack Kiefer*, Vol. 1, 65–78, Wadsworth, Belmont, California.

Kruskal, W. and Majors, R. (1989). Concepts of Relative Importance in Recent Scientific Literature. *American Statistician*, Vol. 43, No. 1, 2–6.

Kuhn, T. (1970/1962). *The Structure of Scientific Revolution*, The University of Chicago Press.

Langley, P. (1981). Data-driven Discovery of Physical Laws. *Cognitive Science*, Vol. 5, 31–54.

Lehmann, E. L. (1985). The Neyman-Pearson Theory after 50 Years. *Proceedings of the Berkeley Conference in Honor of Jerzy Neyman and Jack Kiefer*, Vol. 1, 1–14, Wadsworth, Belmont, California.

Leonard, T. (1983). Some Philosophies of Inference and Modelling. *Scientific Inference, Data Analysis and Robustness*, edited by G. E. P. Box, T. Leonard, and C. F. Wu, 9-23. Academic Press, New York.

Minton, P. D. (1983). The Visibility of Statistics as a Discipline. *The American Statistician*, Vol. 37, No. 4, 284–289.

Morris, C. N. (1987). Comments to "Reconciling Bayesian and Frequentist Evidence" by Casella, G. and Berger, R. L., *JASA*, Vol. 82, 131–133.

Morrison, D.E. and Henkel, R.E. (ed.) (1970). *The Significance Test Controversy*. Aldine, Chicago.

Olshen, R. (1973). The Conditional Level of the F-Test. *JASA*, Vol. 68, 692–698.

Pocock, S. J., Hughes, M. D., and Lee, R. J. (1987). Statistical Problems in the Reporting of Clinical Trials. *The New England Journal of Medicine*, Vol. 317, 426–432.

Trenton Times (1984). Survey: Kids do an Hour of Homework Daily.

Wang, C. W. H. (1986). A Minimum Distance Estimator for First- Order Autoregressive Processes. *Annals of Statistics*, Vol. 14, No. 3, 1180–1193.

Wilkinson, D.T. (1986). Anisotropy of the Cosmic Blackbody Radiation. *Science*, Vol. 232, 1517–1522.

Wonnacott, T.H. and Wonnacott, R. J. (1982). *Introductory Statistics*, 4th edition, Wiley, New York.

Chapter 2

Quasi-Inferential Statistics

I. RANDOMNESS OR CHAOS?

It is interesting to see that many famous statisticians were involved in the discussion of randomness in the years 1985, 1986, and 1987. One lesson we learned from their discussions is that the word "random" is probably the most used and misused in statistical vocabulary.

In the March issue of *Science*, Kolata (1986) described some of Diaconis' interesting work in randomness under the title "What does it mean to be random?" One of Diaconis' conclusions is:

Most phenomena that are thought to be random are not so random after all.

For more discussions along this line, see Diaconis and Engel (1986), and DeGroot (1986).

In statistical inference, if a phenomenon is not random, a man-made randomization can be induced to achieve the randomness needed in a legitimate calculation. But the procedure is not always applicable. Therefore opinions often conflict with each other on the reporting of the "inferential statistics" based on encountered data. For instance, Meier (1986, *JASA*) spelled out that he does not quite share David Freedman's hard-line position against formal statistical inference for observational data. The same attitude was exhibited in

Fienberg (1985), Dempster (1986a), Kadane and Seidenfeld (1987), and Patil (1987).

In a Neyman Lecture held in Chicago, Dempster (1986b) spent some 10 minutes on the following issues:

(A)   chance mechanisms
(B)   personal measures of uncertainty

and argued that

(A) ⊂ (B)

In the February issue of *Statistical Science*, Madansky (1986) commented on Freedman's general quest which includes a never-wavering adherence to randomization. In a conference held in New Jersey, April 1986, Freedman questioned a speaker: "Do you have randomization in your study?" The tone was soft, but the message is serious. The following is a typical example of Freedman's hard-line position (FPP, 1978, pp. 350–351):

> A psychologist administers a test of passivity to 100 students in his class and finds that 20 of them score over 50. He concludes that approximately 20% of all students would score over 50 on this test. He recognizes that this estimate may be off a bit, and he estimates the likely size of the error as follows:
>
> SE for number scoring over 50 $= \sqrt{100 * 0.2 * 0.8} = 4$
>
> SE for percent $= (4/100) * 100\% = 4\%$
>
> What does statistical theory say?

Freedman's answer to the above question is:

> Theory says, watch out for this man. What population is he talking about? Why are his students like a simple random sample from the population? Until he can answer these questions, don't pay much attention to the calculations. He may be using them just to pull the wool over your eyes.

As Cochran (1983) pointed out:

> Experiments in behavioral psychology are often conducted using graduate students, and other volunteer students (paid or unpaid) in a university's psychology department. The target populations may be all young persons in a certain age range.

No wonder many statisticians, e.g., Madansky, Meier and Dempster, expressed their dismay at Freedman's hard-line position. However, on what grounds can the above psychologist calculate SE (Standard Error)? Some attempts to justify the calculation of the standard error of observational data can be found in Freedman and Lane (1983a, 1983b). See also Diaconis (1985b) for a good survey of some theories of data description that do not depend on assumptions such as random samples or stochastic errors.

However, the theories developed by those authors cannot really answer the following question: On what grounds can the above psychologist generalize the result from the grab set to a certain population? This chapter is to provide another view of the issue. For this purpose, we will first present a short discussion of epistemology (the philosophy of science).

II. HUME'S PROBLEM

Scientific thinking includes deductive and inductive reasoning. Deductive reasoning is epitomized by mathematics and logic; the latter deals mainly with validity of arguments rather than the truth of our universe. Inductive reasoning is making a general statement or a scientific law based on certain experimental data. (Note that mathematical induction is neither inductive nor deductive. It is Peano's fifth axiom. Using this axiom to prove theorems is deductive, not inductive. However the mental process of formulating the theorem prior to a formal proof is inductive, not deductive. The name "mathematical induction" is somewhat misleading.)

The inductive principle was first systematically described by Francis Bacon in 1620 and has long been seen as the hallmark of science. However, in 1739, David Hume pointed out that no number of singular observation statements, however large, could logically entail an unrestrictedly general statement. He further argued that *inductive reasoning is of psychology, not of logic*. The trouble with induction, which has been called "Hume's problem," has baffled philosophers from his time to our own (see, e.g., Magee, 1973, pp. 12, 13; Diaconis, 1985a; Russell, 1945; Weaver, 1963; Popper, 1963, 1935; etc.).

Carnap (1937) and many logical positivists proposed the degree of confirmation for scientific theory and thus tried to formulate an inductive logic. This attempt appears sensible but proved to be exceptionally difficult. Indeed, no inductive logic seemed capable of capturing man's intuitions about what confirms a theory and what does not.

When a man cannot explain a certain natural phenomenon, he usually creates new terminology to satisfy himself. According to Popper (1974, p. 1015), inductive logic is one of those terminologies—it is only an illusion; it does not exist. Sir Karl Popper (1963) asserted that "the way in which knowledge progresses, and especially our scientific knowledge, is by unjustified (and unjustifiable) anticipations, by guesses, by tentative solutions to our problems, by conjectures." He also argued that those conjectures "can neither be established as certainly true nor even as 'probable' (in the sense of the probability calculus)."

Russell (1945, p. 674) wrote: "Induction is an independent logical principle, incapable of being inferred from either experience or from other logical

principle, and that without this principle science is impossible." It is sometimes said that induction is the glory of science and the scandal of philosophy.

In his classic book *Lady Luck*, Warren Weaver (1963, p. 308) linked statistics to inductive logic: "Francis Bacon was the first properly to emphasize
inductive methods as the basis of scientific procedure, but it was not until 1763
that the English clergyman Thomas Bayes gave the first *mathematical basis to
this branch of logic* " [emphasis supplied].

However, if induction is "a branch of logic," it must be a special kind of
"fuzzy logic."[1] This "fuzzy logic" includes cognition, theory formulation,
generalization, model fitting, and causal inference. Apparently the statistical
discipline has close ties and great contributions to inductive reasoning. However, the practice of statistical reasoning is not always so straight-forward as
presented in many popular textbooks. As a matter of fact, the whole notion of
"statistical inference" often is more of a plague and less of a blessing to
research workers.

In contrast to non-scientific statements (which are based on authority, tradition, emotion, etc.), scientific statements are induced or deduced from facts
and empirical evidences. A big problem is that many so-called "empirical evidences" are not evidence at all. For example, see Figs. 1 and 2. (Luckiesh,
1965): In Fig. 1, both lines are of equal length. In Fig. 2 "twisted cords" are
indeed concentric circles. "Evidence" of this sort is plentiful in human reasoning, especially in causal inferences (and Bayesian analysis as well). For examples in this regard, see Chapters 3, 4, and 5.

III. UNOBSERVABLES, SEMI-UNOBSERVABLES, AND GRAB SETS

Statistical inference, which generalizes results from a sample to a population,
is as confusing and scandalous as induction. In fact statistics is often classified
as a subcategory of lies (Moore, 1985). To earn our credit back, let's first consider the following question: Is statistical inference deductive or inductive?

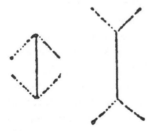

Figure 1   Which line is longer? Copyright © 1965 by Dover Publications, Inc.
Reprinted by permission.

Figure 2   "Twisted cord." Copyright © 1965 by Dover Publications, Inc. Reprinted by permission.

Given a fixed population, the answer is simple: (1) If we don't have randomization, then the generalization is inductive. (2) If we do have randomization, then the inference is deductive.

In a sample survey, the very act of randomization (mixing the tickets in a box) completely changes the nature of "the unobserved." Indeed, after randomization, "the unobserved" are only "semi-unobserved," because they have been stirred by the act of randomization and certain characteristics of the population units can now be estimated with desired accuracy. The generalization is then based on Kolmogorov's probability theory, which is of logic, not of psychology.

The summary statistics drawn from a grab set are of great value for scientific endeavors, but they are descriptive, not inferential statistics. Many people interpret the data of a grab set as a random sample from some hypothetical universe composed of data elements like those at hand. Put another way, the sample is used to define the population. Such a hypothetical population is only conceptual—it may not exist (see Henkel, 1976, p. 85).[2]

To illustrate, let's look at an example on the scientific sampling of seat-belt usage which recently became mandatory in New Jersey. In a small experiment, this author took classroom polls to estimate the percentage of seat-belt usage in New Jersey. The percentages from five different classes are as follows (the polls were taken from April to June, 1986):[3]

Class 1	15/23	=	65%
Class 2	10/20	=	50%
Class 3	16/19	=	84.2%
Class 4	9/9	=	100%
Class 5	10/23	=	43.5%

In March, 1985, a college student who worked for a sociology professor obtained 457/750 = 61%; the sampling method used was by counting at one entrance of the college. It is interesting to compare the results reported by two local newspapers, the *Princeton Packet* and the *Trenton Times*. The *Princeton Packet* (December 17, 1985) reported 40%; sample size and sampling method were not indicated. The *Trenton Times* (December 8, 1985) reported 32.35% for drivers aged 26 and under in New Jersey, but 19% in Pennsylvania where seat-belt usage was not mandatory. The sampling method of the *Trenton Times* was also not indicated. In March 23, 1986, the *Trenton Times* reported that 50% of people always wear seat belts in New Jersey and 67% of people wear seat belts always or most of the time.

For cause-and-effect, the *Trenton Times* (December 17, 1985) reported that a 30% drop in front-seat auto fatalities on New Jersey highways during October 1985 is attributed to the seat-belt law. On the same issue, the *Princeton Packet* reported a 10.6% decline from March to October. After comparisons of the above results, the students were asked:

1. Which of the above samples are drawn by scientific sampling?
2. Which of the above samples can be used to construct a 95% confidence interval of the true percentage?

Scientific results in respected journals and statistics books are expected to be more reliable than ordinary newspaper results. However, this reliability is often questionable. It is very difficult to stop people from calculting SE out of grab sets once they have learned (or thought they had learned) the formulas. The situation is even worse if they have easy access to computer packages (SAS, SPSSX, MINITAB, etc.) which provide inferential statistics for any grab set. In a book review I wrote for a publisher, I was surprised that two prominent statisticians were calculating inferential statistics from a grab set; they didn't even bother to ASSUME the randomness of the data. With great dismay, I wrote:

> The authors' confidence (at 95% level) is based on their personal psychological belief, not on statistical theory.

This psychological indulgence in calculating confidence intervals on grab sets is not uncommon. Similar examples can be found almost everywhere. This brings up a serious question.

IV. IS STATISTICS A SCIENCE?

The answer is YES and NO. The deductive part of statistics (mathematical statistics, probability) is exact science (in the sense that it is internally consistent). Without this, applied statistics is only an act of personal belief. The inductive part of statistics can be good science, but can also be pseudo-science hiding behind a facade of irrelevant computations.

Sir Ronald A. Fisher (quoted by FPP, 1978, p.10) once said:

That's not an experiment you have there, that's an experience.

Statisticians who are so casual in calculating 95% confidence intervals from grab sets should think twice about Fisher's advice. Freedman and Navidi (1986, p. 39) wrote: "Like all academics, we insist on making a distinction." It is this stand which is consistent with their training in statistical theory; and it is this stand which qualifies some statisticians as truer scientists than those who draw inference from any grab set.

In one example, Thomas Reider (1976, p. 73), an astrologist, calculated probability and standard deviations to demonstrate "a startling correlation between planetary circles and stock market patterns!" If we do not insist on making a distinction, how can we tell a statistician from an astrologist?

The statistical enterprise is a complex and often erratic one. It includes many diversified fields and a large number of people with different interests, skills, and depths of understanding. Many respectable mathematical statisticians devote their lives to the foundation of mathematical statistics and strictly keep their hands off data. However, the majority of IMS (Institute of Mathematical Statistics) and ASA (American Statistical Association) members are involved in consulting, teaching, or writing statistics books. It is not a bad thing to see theoretical statisticians working on applied problems. But a caveat to those statisticians is this: There is often a big gap between statistical theory and reality.

For instance, it is very difficult to construct a non-measurable set on the real line (Royden, 1968), but in the real world many events are simply not "measurable" or can only be poorly measured. For example, how is one going to measure a student's progress in a philosophy course? Furthermore, assume that there is a good measure for this purpose, can one apply the Kolmogorov Extension Theorem (Chow and Teicher, 1978, pp. 185–186) to construct an infinite sequence of i. i. d. (independent and identically distributed) random variables and to establish a "scientific" measure of the student's progress?

The above "application" of measurable functions and of Kolmogorov Extension Theorem may appear silly; but looking at the current state of statistical practice, for instance a field called survival analysis, many applications of the theory are similarly far-fetched. In fact, in a semi-public setting, a leading statistician asserted that "the whole survival analysis is ridiculous!" We have no intention of endorsing such a statement at this moment, but we do feel that many statisticians came to this field simply dumping their useless research into scientific literature, and that a thorough review of the current practice of survival analysis will help the reputation of our profession.

It is disheartening to see that many theoreticians totally lost their rigorous training in the assumptions and solid proofs when they approached real-life data. The principle of randomization is just one of the things they seem to forget or ignore. This attitude toward applying statistics is ethically questionable and is in fact counterproductive to the goal of science. Freedman has long been railing against this inattention. It is good to see that he is getting echoes from different valleys.

Many social scientists have been using statistics intensively because their discipline "has a problem legitimizing itself as a science" (Nell Smelser of U.C. Berkeley; quoted from the *New York Times*, April 28, 1985, p. E7). Sociologist William Sewell (1985, also in the *New York Times*, April 28, 1985, p. E7) of the University of Wisconsin asserted that "the mainstream of sociology is empirical and quantitative" and advocated large-scale computers and large-sample surveys. Quantitative psychologists, biologists, medical researchers, and economists are all using statistics as often as sociologists. However, a question is: "Are current practices of statistical methods scientific or not?" Our answer is more on the negative side and the examples in this regard will continue to unfold.

Freedman, as apparent from many of his writings, seems determined to confront the gross misuse of statistical methods. Madansky (1986) and others worry that Freedman's quest may put most statisticians out of business. Madansky dubbed Freedman the "neturei karta" (a Hebrew phrase, meaning the purist or the guardian of the city) of statistics, and concluded that if Freedman is successful, then "the State of Israel" (statistics) will cease to exist.

Like Madansky, we were once troubled by Freedman's hard-line position until philosophy of science came to the rescue. In the following section, we will try to explain this philosophy and to further popularize Freedman's position.

V. GRAB SETS AND QUASI-INFERENTIAL STATISTICS

To experienced data analysts, the characteristics of real-life data usually are like this: non-normal, non-random, heavy prior, and small sample. Data of this type are of great value for intellectual exploration. However, the whole practice of "statistical inference" based on this kind of data is problematic and in many ways deceiving. This conclusion is not new (see, e.g., Selvin, 1957 and FPP, 1978), yet the practice remains dominant and fallacious in empirical science.

As suggested in FPP (1978, p. 407), given a chunk of data scientists might be better off to report only descriptive, not inferential, statistics. But this suggestion is too difficult to swallow for most statisticians and empirical scientists. Perhaps, those reported standard errors should be called "Quasi- inferential statistics." This is a modification of Freedman's hard-line position: it is intended to utilize those statistics as a supplement to Tukey's EDA (Exploratory Data Analysis); it is also intended to popularize Freedman's position— i.e., to draw a sharp line between formal and casual statistics.

For example, Efron and Tibshirani (1986, p. 67) reported that the Pearson correlation coefficient for 15 LSAT-GPA data points, each corresponding to an American law school, is .776 with a bootstrap confidence interval (.587, .965). This calculated confidence interval should not be taken literally as an inferential statistic for the following reasons: (1) Randomness of the data is questionable. (2) Inference based on aggregated data, such as correlation of law school data, is quite often misleading and may easily get into hidden traps (see examples in FPP, 1978, p. 14, pp. 141–142, etc.). In other words, from Efron and Tibshirani's calculation, it appears that there is a correlation between LSAT and GPA; however, to quantify this belief within the range (.587, .965) with 90% confidence is psychological, not statistical.

The calculated confidence interval or P-value should be billed "quasi-inferential statistic" to serve another purpose in EDA: *If the P-value or the width of the confidence interval is significant, then the descriptive statistics are useful for intellectual speculations; otherwise, a larger sample (or a prior distribution) may be needed.* A similar interpretation was provided by Paul Meier (1986):

> If the observed association would not be counted statistically significant had it arisen from a randomized study, it could not be counted as persuasive, when even that foundation is lacking. If the observed association is highly statistically significant, however, the extent of its persuasiveness depends on many uncertain judgments about background factors, and its persuasive value is not at all reflected in the significance level itself.

The next example demonstrates the use of quasi-inferential statistics in a chi-square test. In a clinical social science study, three groups of scientists were asked to answer YES or NO on a specific question and yielded the following two-way table:

	Yes	No	
Group 1	68.8%	31.2%	100%
Group 2	91.3%	8.7%	100%
Group 3	80.65%	19.35%	100%

In comparison of 31.2% and 8.7% (the extreme values in the second column), the difference seems practically significant. However, the computed P-value of chi-square test (based on the original raw count) is .404, which is not significant statistically. Because the original data are from a grab set, the calculation of the P-value "does not make sense"—according to FPP (1978, p. 497). However, the P-value was obtained automatically from a handy software package, and it makes perfect sense to tell a client that a larger sample or some prior knowledge of the subject is needed in order to draw a more persuasive conclusion.

A harder problem is: If the above grab set is large enough to produce a small P-value, then what should we say to that investigator who is desperate to find a significant P-value to support his result? Should we reject the project in the first place? The answer is NO. At least those "quasi-inferential statistics" are useful for exploratory analysis. This assertion can be better illustrated by the following example.

Jonathan has to reach the airport within 35 minutes. Suppose that in the past he had tried two different routes. On the first pair of trials, he spent 32 minutes on route 1 but less on route 2 (27 minutes). On the second pair of trials, the time on route 1 increased to 32.25 minutes, while that on route 2 decreased to 19 minutes. The data can be summarized as follows.

	Route 1	Route 2
First trial	32	27 min
Second trial	32.25	19 min

On the basis of this information, should Jonathan take route 1 or 2 in order to catch the airplane within 35 minutes?

Given the data, many students in my class replied with the flat answer "route 2," because it took less time. But if we assume the normality of the data, then the P-values of the t-tests (H_o: $\mu \geqslant 35$) are 2.8% for route 1 and 20.5% for route 2, respectively. Therefore, it is route 1 which is significantly preferable, not route 2. But this conclusion is against our intuition. Which one should we trust: the mysterious statistical test or our intuition?

The 95%-confidence intervals work out (30.5, 33.7) for route 1 and (−27.8, 73.8) for route 2, respectively; they provide convincing evidence that route 1 would be a better bet.[4] But then some students challenged the instructor: "How can we have negative time for route 2?"

The answer relates to the normality assumption of the travel times. Further, why does route 2 have bigger variability than route 1? The issue now narrows down to the chance mechanism which generated the data.

The fact is that route 1 is highway drive while route 2 goes through downtown traffic which is where the variability came from. The normality assumption is reasonable for route 1 but not for route 2. One more reason for choosing route 1 is that one can speed up a little bit, to say 60 mph, but there is not much one can do on route 2.[5]

The next example of quasi-inferential statistics was inspired by a regression analysis in Larson (1982, p. 463):

X: 300, 351, 355, 421,
Y: 2.00, 2.70, 2.72, 2.69,
X: 422, 434, 448, 471, 490, 528
Y: 2.98, 3.09, 2.71, 3.20, 2.94, 3.73

By standard calculation, $\hat{a} = .557$, $\hat{b} = .00549$. Since $\hat{a}$ is very large and $\hat{b}$ is extremely small, a person without statistical training might reject H_o: $a = 0$ and accept H_o: $b = 0$. However, the 90% confidence intervals (Larson, p. 484) are $(-.364, 1.478)$ for a and $(.00333, .00765)$ for b. Hence, statistically $\hat{a}$ is not significantly different from 0, but $\hat{b}$ is. A conclusion from the above quasi-inferential statistics is this: X and Y are strongly correlated, but the data do not give a good estimate of the intercept of the regression line.

Data analysis can be more revealing if the background information is provided. In this case, X = entrance exam scores and Y = graduating GPA scores for 10 recent graduates of a certain university. This piece of information plus common sense and the high R-square value (the so-called percentage of explained variability, which is 74% in this case) cast doubt on our previous belief that the estimated $\hat{b}$ is reliable. A graphic display of the data is shown in Fig. 3. Cover the two extreme points at X = 300 and 528 and then see what's

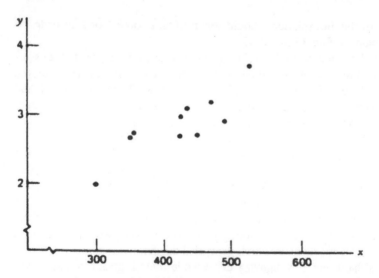

Figure 3   Grade-point average vs. entrance exam scores. Copyright © 1982 by John Wiley & Sons, Inc. Reprinted by permission.

left. If those two points are deleted, then the new estimation of b ($\hat{b} = .0024$) is barely significant ($P = 12\%$ and the Bayes factor $= P\sqrt{n/100} = 3.4\%$; see Good, 1987). The decision to delete the two extreme points is also supported by Cook's (1979) D statistics which are used to measure the influence of each observation on the estimates:

X: 300, 351, 355, 421,
D: .515, .138, .124, .035,
X: 422, 434, 448, 471, 490, 528
D: .011, .024, .121, .006, .269, .550

Note that the D statistics are overwhelming at $X = 300$ and 528. After we delete the outliers, the size of $\hat{b}$ (0.0024 vs. 0.00549, the original $\hat{b}$) and the R-square (35% vs. 74%, the original R-square) reinforce our common sense that among average students in the *same* university, the entrance exam score may be only weakly correlated with their GPA.

The demonstration of quasi-inferential statistics in the above examples is intended to legalize the calculation of standard errors for any grab set. In certain situations, the quasi-inferential statistics serve surprisingly well. This happens when the target population is relatively homogeneous. However, the belief in a homogeneous population is largely subjective. The basis of this belief is extra-statistical, in the sense that it is inductive, not deductive.

VI. CONCLUDING REMARKS: QUASI- AND PSEUDO-INFERENTIAL STATISTICS

The demarcation between genuine- and quasi-inferential statistics is a *clearly defined and well-justified chance model*. In many statistical applications (especially in sample survey and clinical trials), this means randomization. However, any scientific tool has its limitation; randomization is no exception (see examples in FPP, 1978, p. 311–315). Furthermore, it is a simple fact that the majority of data are taken under non-randomized conditions. In these kinds of situations, a combination of EDA, quasi-inferential statistics, and a healthy curiosity about data may help to shed light on many complicated problems. But we have to spell out the limitations of the reported numbers, and the final interpretation should be left to subject-matter experts.

It is important to note that without randomization, statistical inference can easily go off the rails (see examples in FPP, 1978, Chapters 1, 2, 19, etc.). When drawing a generalization from a grab set, the nature of population has to be carefully examined. Otherwise, those quasi-inferential statistics will quickly deteriorate into "pseudo-inferential statistics"—they fool both the general public *and* the investigators as well.

NOTES

1. Some computer scientists use terms like "fuzzy logic," "fuzzy mathematics," etc. (see, e.g., Shimura, 1975). In this article, the term "fuzzy logic" is used in a non-fuzzy manner: it means "loose logic" which is not logic at all.
2. Freedman and Lane (1983a, pp. 185, 198–199) pointed out that the notion of this hypothetical universe is due to R. A. Fisher and its first criticism is due to E. T. Pitman.
3. Class 1: Junior and senior students (all aged 26 and under); class 2 and 5: mostly sophomores and juniors; class 3: graduate students; class 4: faculty members.
4. Significance tests in this example are used only as a mathematical exercise. The tests are redundant in this specific case.
5. This example is indeed a modern version of Fisher's puzzle (1934, pp. 123–124). It is important to note that all the "quasi-inferential statistics" in this example do not have the same solid footing as the ones in a randomized controlled experiment: the calculated statistics in this travel problem can easily be overridden by other background knowledge of the data.

REFERENCES

Carnap, R. (1937). Testability and Meaning. *Philosophy of Science*, Vol. 3, 419–471, and Vol. 4, 7–40.

Chow, Y. S. and Teicher, H. (1978). *Probability Theory*. Springer-Verlag, New York.

Cochran, W. (1983). *Planning and Analysis of Observational Studies*. Wiley, New York.

DeGroot, M. H. (1986). A Conversation with Persi Diaconis. *Statistical Science*, Vol. 1, No. 3, 319–334.

Dempster, A. P. (1986a). Comment on Regression Models for Adjusting the 1980 Census. *Statistical Science*, Vol. 1, No. 1, 21–23.

Dempster, A. P. (1986b). Employment Discrimination and Statistical Science, a Neyman Lecture in the ASA Annual Meeting in Chicago. In *Statistical Science*, 1988, Vol. 3, No. 2, 149–195.

Diaconis, P. (1985a). Bayesian Statistics as Honest Work. *Proceedings of the Berkeley Conference in Honor of Jerzy Neyman and Jack Kiefer*, Vol. 1, 53–64, Wadsworth, Belmont, California.

Diaconis, P. (1985b). Theory of Data Analysis: From Magical Thinking Through Classical Statistics. *Exploring Data Tables, Trends and Shapes*, edited by D. Hoaglin, F. Mosteller and J. Tukey. 1–36. Wiley, New York.

Diaconis, P. and Engel, E. (1986). Comment on Some Statistical Applications of Poisson's Work. *Statistical Science*, Vol. 1, No. 2, 171–174.

Efron, B. and Tibshirani, R. (1986). Bootstrap Methods for Standard Errors, Confidence Intervals, and other Measures of Statistical Accuracy. *Statistical Science*, Vol. 1, No. 1, 54–77.

Fienberg, S. E. (1985). Comments and Reactions to Freedman, Statistics and the Scientific Method. *Cohort Analysis in Social Research*, edited by W. M. Mason and S. E. Fienberg. Springer-Verlag, New York.

Fisher, R. A. (1934). *Statistical Methods for Research Workers*, 5th edition. Edinburgh, Oliver and Boyd.

Freedman, D. A. and Lane, D. (1983a). Significance Testing in a Non-stochastic Setting. *A Festschrift for Erich L. Lehmann*, edited by P. J. Bickel, K. A. Docksum and J. L. Hodges, Jr. 185–208. Wadsworth International Group, Wadsworth, Belmont, California.

Freedman, D. A. and Lane, D. (1983b). A Non-Stochastic Interpretation of Reported Significance Levels. *J. of Business and Economic Statistics*, Vol. 1, 292–298.

Freedman, D. A., Pisani, R., and Purves R. (1978, 1980). *Statistics*. Norton, New York.

Good, I. J. (1987). Comments to "Reconciling Bayesian and Frequentist Evidence" by Casella, G. and Berger, R. L., *JASA*, Vol. 82, 125–128.

Henkel, R. E. (1976). *Tests of Significance*. Sage, Beverly Hills, California.

Kadane, J. and Seidenfeld, T. (1987) Randomization in a Bayesian Perspective. Presented in "Foundations and Philosophy of Probability and Statistics," an International Symposium in Honor of I. J. Good, Blacksburg, Virginia.

Kolata, G. (1986). What does it Mean to be Random? *Science*, Vol. 231, 1068–1070.

Larson, H.J. (1982). *Introduction to Probability Theory and Statistical Inference*, 3rd edition. Wiley, New York.

Luchiesh, M. (1965). *Visual Illusions*. Dover, Mineola, New York.

Madansky, A. (1986). Comment on Regression Models for Adjusting 1980 Census. *Statistical Science*, 28–30.

Magee, B. (1973). *Karl Popper*. Viking Press, New York.

Meier, P. (1986). Damned Liars and Expert Witness. *JASA*, Vol. 81, No. 394, 268–276.

Moore, D. S. (1985). *Statistics: Concepts and Controversies*. W. H. Freeman and Company, San Francisco, California.

Popper, K. R. (1935,1959). *The Logic of Scientific Discovery*. Harper and Row, New York.

Popper, K. R. (1963). *Conjectures and Refutations: The Growth of Scientific Knowledge*. Harper and Row, New York.

Popper, K. R. (1974). Replies to My Critics. *The Philosophy of Karl Popper*, edited by P. A. Schilpp. The Library of Living Philosophers, Open Court, La Salle, Illinois.

Reider, T. (1976). *Astrological Warnings and the Stock Market*. Pagurian Press, Toronto, Canada.

Royden, H. L. (1968). *Real Analysis*, 2nd edition. Macmillan, New York.

Russell, B. (1945). *The History of Western Philosophy*. Simon and Schuster, New York.

Selvin, H. (1957). A Critique of Tests of Significance in Survey Research. *American Sociological Review*, Vol. 22, 519–527.

Shimura, M. (1975). An Approach to Pattern Recognition and Associative Memories using Fuzzy Logic. *Fuzzy Sets and their Applications to Cognitive and Decision Processes*. Academic Press, New York.

Weaver, W. (1963). *Lady Luck: The Theory of Probability*, Doubleday, Garden City, New York.

Chapter 3

Statistical Causality and Law-Like Relationships

I. INTRODUCTION

In empirical science and decision-making, inferences regarding cause-and-effect occur all the time. Statistical tools, as one can expect, provide a big boost to a causal argument. As a scientific tool, however, statistical reasoning offers a mixed blessing to inferences about causes and effects. One reason is that statistical causality toes a thin line between deduction and induction, and often fuels confusion on the part of researchers and decision-makers.

This chapter is thus devoted to a discussion of the use and misuse of statistics in causal inference. It begins, in Section II, with several examples of abuses and compares orthodox statisticians' view with causal inferences drawn by lay practioners.

In Sections III–IV, the chapter looks at some scholarly exchanges on the subject of causality. Strictly speaking, the term "probabilistic causality" is self-contradictory. As a consequence, debates on the foundation of probabilistic causality are sporadic in the literature. Recently, in his paper entitled "Statistics and Causal Inference," Holland (1987, JASA) formulated the "Fundamental Problem of Causal Inference" and stated the motto: *"No Causation without Manipulation."* Holland's article has since drawn attention (and criticism) from other respected authorities.

With Holland's motto in mind, the discussions in Sections III and IV delineate the boundary between "controlled experiments" and "observational studies," a distinction which seems obvious but is often hard to maintain. Section V also discusses the problems of drawing causal inference in sample surveys. (Certain portions of Sections III and IV are rather technical. Some readers may prefer to skip these parts in their first reading.)

In sharp contrast to Holland's motto is the position taken in the book, *Discovering Causal Structures: Artificial Intelligence, Philosophy of Science, and Statistical Modeling* (Glymour et al., 1987). Glymour advocates the use of statistical models for nonexperimental science; the project was funded by three National Science Foundation grants and will be briefly reviewed in Section VI. A case study is presented in Chapter 4 to demonstrate the potential defects in Glymour's ambitious program.

II. SENSE AND NONSENSE IN CAUSAL INFERENCE: EXAMPLES

EXAMPLE 1   A Harvard study (*The New York Times* and the *Trenton Times*, February 10, 1988), conducted by Dean K. Whitla, director of instructional research and evaluation, concluded that Harvard students who took courses to prepare themselves for the SAT (Scholastic Achievement Tests) scored lower than students who did not take the classes. The study concluded that SAT coaching provides no help, and that "the coaching industry is playing on parental anxiety" (Harvard University Admissions Director, W. Fitzsimmons).

The study was based on questionnaires distributed to Harvard freshmen in Fall 1987. Of the 1409 participants, the 69% who had taken no coaching classes scored an average of 649 on the verbal portion of the SAT and 685 on the math. Students who had been coached (14% of the 1409 participants) scored an average of 611 on the verbal and 660 on the math:

	No coaching	Coached	Difference
VSAT	649	611	+38
MSAT	685	660	+25

To the uninitiated, the data definitely support the claims of the Harvard officials. But on the second look, the study was obviously flawed; you don't need a statistician to figure out that good students do not need coaching and will still score higher than poorer students who were coached. In a similar situation, people who go to hospitals are in general less healthy than those who don't go to hospitals; but this fact does not imply that going to hospitals tends to weaken a person's health condition.

It appears that the Harvard officials' original intention to discourage students from going to coaching school is admirable. But the statistics they used to promote their argument simply do not bear scrutiny.

EXAMPLE 2   In the fall of 1987, a study was conducted by the American Council on Education and the Higher Education Research Institute at the University of California, Los Angeles. The study involved 290,000 college freshmen and found that a record number (75.6%) identified the goal of "being very well off financially" over the goal of "developing a meaningful philosophy of life" as "essential" to them, or "very important." This is up from 73.2% one year earlier, 70.9% in 1985, and nearly double the level in 1970, 39.1%. Fig. 1 shows the graphic representation of these numbers.

The survey was done scientifically. But how to interpret the findings? In a news story, *The New York Times* (January 14, 1988) concluded that there is a rising "trend of materialism." This conclusion seems to be valid, but Professor Merrill Young of Lawrence University argued otherwise (*The New York Times*, February 19, 1988).

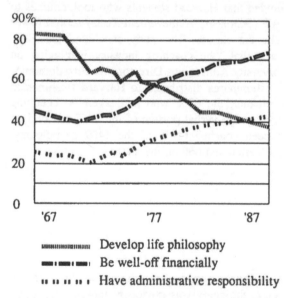

||||||||||||||||||||| Develop life philosophy
■■■■■■■■ Be well-off financially
'' '' '' '' '' ' Have administrative responsibility

Figure 1   College Student's Lifetime Goals. Responses of 290,000 college freshmen asked to list, in order of priority, the following: developing a life philosophy; being well off financially; having administrative responsibility. (Source: UCLA. Survey conducted annually among college freshmen since 1967.) Copyright © 1992 by Higher Education Research Institute, UCLA. Reprinted by permission.

According to Professor Young, money was of little concern to the students before 1977, because they considered that if they graduated, they would find it "practically lying in the street." In contrast, since 1977 Professor Young has witnessed increasing doubt on the part of students about their chances of getting jobs after graduation. He concluded that "the rising trend has not been in materialism, but in pessimism."

Professor Young also attributes most comments about the greed of today's college students to the now middle-aged members of the censorious college generation of the late 1960s and early 1970s. "They spent their 20s scolding their elders," the Professor complained. "Now they are spending their 40s lecturing the young."

Reading this penetrating analysis, one may speculate still further. First, why does it have to be *either* materialism *or* pessimism? Isn't it possible that *both* factors are responsible for the new trend, in a dynamic fashion? Second, what happened in 1977? Can we learn something about this specific year and use the information in a constructive way? So far no definitive answers to these questions appear to be available.

EXAMPLE 3   In an attempt to help people assess the quality of health care, the U.S. Government issued data showing the mortality rate at nearly 6,000 hospitals (*The New York Times*, December 18, 1987). The new statistics, contained in seven large books, were derived from data on 10 million hospital admissions and 735,000 deaths among Medicare beneficiaries in 1986. The lesson of the study was this: Watch out if the mortality rate of a hospital is higher than expected.

But is mortality rate a relevant indicator? If one wants to know the quality of health care, there are many obvious factors to be considered: the equipment in a hospital, the staff, the environment, the infectious control program, and the quality control procedures (nursing, pharmacy, standardization, etc.).

In addition mortality rates are deeply confounded by the fact that good hospitals tend to attract the severely ill, who in turn inflate a hospital's mortality rate. Also, hospitals are known to send people home to die to keep them off the mortality list. Thus, an alternative conclusion might be: Watch out if the mortality rate of a hospital is *lower* than expected.

EXAMPLE 4   In this example we discuss the effect of statistical training on reasoning. It is believed by certain behavioral scientists that the teaching of statistical rules in an abstract way, as it is normally done in statistics courses, has an effect on the way people reason about problems in everyday life (Holland, Holyook, Nisbett, and Thagard, 1987, p. 256). The scientists cited a study by Fong et al. (1986; funded by Grants from NSF, SES, ONR, etc.). A similar study by Fong et al. (1987) was published in *Science* and was cited in *Statistical Science* as good evidence that "the study of statistics, as taught by psychologists, improves students' ability to deal with uncertainty and variability in quite general settings" (Moore, 1988).

The Fong (1986) study examined four groups of subjects differing widely in statistical training: (1) college students who had no statistical training at all, (2) college students with one or more statistics courses, (3) graduate students in psychology, most of whom had had two or more semesters of statistical training, and (4) Ph.D.-level scientists who had had several years of training.

Subjects were presented with a problem about restaurant quality. There were two versions: randomness and no randomness. Within each group tested, half of the subjects received the cue and half did not.

In the no-randomness cue version, a traveling businesswoman often returns to restaurants where she had an excellent meal on her first visit. However, she is usually disapppointed because subsequent meals are rarely as good as the first. Subjects were asked to explain, in writing, why this happened.

In the randomness cue version, an American businessman traveling in Japan did not know how to read the menu; so he selected a meal by blindly dropping a pencil on the menu and ordering the dish closest to it. As in the other version, he is usually disappointed with his subsequent meals at restaurants he originally thought were superb. Why is this?

Answers were coded into two categories: (1) nonstatistical, if the subject assumed that the initial good experience was a reliable indicator that the restaurant was truly outstanding, and attributed the later disappointment to a definite cause (e.g., "Maybe the chef quit"); and (2) statistical, if the subject suggested that meal quality on any single visit might not be a reliable indicator of the restaurant's overall quality (e.g., "odds are that she was just lucky the first time"). The results were summarized in Fig. 2 (where $p = \%$ of statistical responses). Fong (p. 277) concluded that

> the chart demonstrates clearly that the frequency of statistical answers increased *dramatically* with the level of statistical training, chi-square(6) $= 35.5$, $p < .001$. [emphasis supplied].

As a statistics educator, I am unable to be so optimistic in the interpretations of the chart: (1) If group 1 "NO STATISTICS" is relabeled as "NO COMPUTER SCIENCE," and groups 2–4 are reclassified accordingly, then we get Fig. 3. A statement now can be drawn in parallel to Fong's conclusion: the above chart demonstrates clearly that the frequency of statistical answers increased dramatically with level of *computer science* training, chi-square(6) $= 35.5$, $p < .001$. The same trick might work if we replace "computer science training" with "art and humanity," or "English and history," etc. In other words, the chart proves nothing except that when one gets older, he or she tends to be more aware of the variability in real life. (2) It is a shame that 80% of group 2 (who had had one or more statistics courses) and 60% of group 3 (graduate students) failed to give statistical responses if they didn't

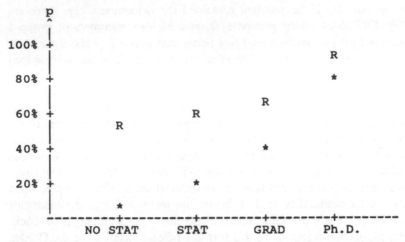

Figure 2 Percent statistical response classified by the levels of subjects' statistical training. R: Randomness cue; * No randomness cue.

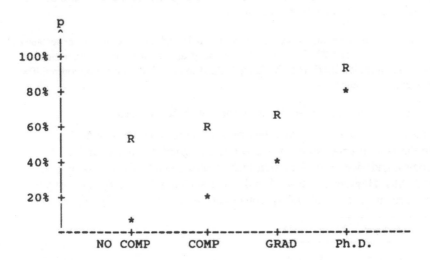

Figure 3 Percent statistical response classified by the levels of subjects' computer training. No COMP = No computer science; COMP = computer science.

receive the cue. (3) If the problem contained the randomness cue, there are only tiny differences among groups 1, 2, and 3. Also, members of group 1 (who received the probabilistic cue) fare better than group 3 (without the cue): 50% vs. 40%. Evidently years of training on statistical formulas were wiped out by a single cue. (4) Twenty percent of Ph.D. scientists failed to think statistically in this study, and the percentage may be much higher in everyday life.

Reasoning in everyday life is, in my observation, almost independent of (or negatively correlated with) a person's academic achievement. On numerous occasions I have witnessed academicians (myself included) drawing ludicrous conclusions if the subject-matter is outside their technical focus. The situation can easily turn ugly if ego, emotion, or self-interest set in. For example, when students fail an exam, they tend to blame the instructor, and the instructor (often a Ph.D.) tends to blame the students for their laziness, poor backgrounds, etc. Causal inference of this sort has repeated itself since the Garden of Eden.

In a more believable study, behavioral researchers at the University of Pennsylvania found that people tend to judge decision-makers on the basis of factors that have more to do with luck than ability (Baron and Hershey, 1988). The researchers interviewed a group of students and asked them to rate hypothetical decision-makers according to variable outcomes. For instance, case 1 read as follows:

A 55-year-old man had a heart condition. He had to stop working because of chest pain. He enjoyed his work and did not want to stop. His pain also interfered with other things, such as travel and recreation.

A type of bypass operation would relieve his pain and increase his life expectancy from age 65 to age 70. However, 8% of the people who have this operation die from the operation itself. His physician decided to go ahead with the operation. The operation succeeded.

Evaluate the physicians's decision to go ahead with the operation.

Case 2 was the same except that the operation failed and the man died.

In the face of uncertainty, it is obvious that good decisions can lead to bad outcomes, and vice versa. Decision makers thus cannot infallibly be graded by their results. However, in cases 1 and 2, the student ratings reflect the opposite of this reasoning (n = 20 undergraduates at the University of Pennsylvania):

Case	Outcome	Mean	SD
1	Success	0.85	1.62
2	Failure	− .05	1.77

Out of 140 pairs of similar cases that differed only in success or failure, higher ratings were given by students to the cases with success in 62 pairs and with failure in 13 pairs. In sum, the study found that the students show an outcome bias, even though they think they should not (or even though they think they do not). The study also found that this tendency colors decisions in law, politics, regulation, institutions, and nearly every other facet of human life.

•   •   •   •   •   •   •   •   •

In certain branches of science, percentages (and statistical jargon) alone appear to be enough to establish the credibility of a causal link. In such disciplines, nothing can be perfect (as argued by some), therefore everything goes. To clean up certain confusions, the next two examples are designed to illustrate the origin and content of hard-line statisticians' views on statistical causality. (This kind of statistician is becoming, in my judgment, an endangered species.)

EXAMPLE 5   In the mid-1950s, the standard method for the treatment of stomach ulcers involved surgery. In 1958, a researcher developed a new method to treat stomach ulcers, not involving surgical intervention. He applied the method to 24 patients, and all were reported cured. The success rate was 100%.

If your relatives or close friends had stomach ulcers, would you recommend the new treatment? (In a recent class, all of my students answered yes.)

In order to draw a causal inference, an experienced statistician first considers how the data were generated. In this case, the technique used for treating ulcers is called gastric freezing (FPP, 1978, p. 7). The procedure goes like this: A balloon is placed in a patient's stomach, and coolant is pumped through the balloon; this treatment freezes the stomach, stops the digestive process, and the ulcer begins to heal.

After you know how the data were generated, do you still recommend the new treatment? (At this point all of my students say no.)

In 1963, a double-blind randomized controlled experiment was conducted to evaluate gastric freezing. There were 82 patients in the treatment group and 78 in the control group, *assigned at random*. The conclusion based on this controlled experiment was that gastric freezing was worthless (see details in FPP, p. 8). The balloon method eventually slipped into the black hole of history.

In scientific endeavors, the task of drawing conclusions about cause-and-effect is often confounded by unexpected lurking variables. Powerful techniques for ironing out the collective effect of confounding factors were developed by Sir Ronald A. Fisher. The techniques were used by scientists in the evaluation of gastric freezing and now are widely used in agriculture, biomedical research, and engineering applications.

The contribution of R. A. Fisher was the origin of a new branch of science, now called statistics. The foundations of the Fisherian tradition are not only mathematical formulas but *also* the physical act of randomization. This tradition thus advanced from medieval statistics (which was solely based on enumeration) to a branch of modern science.

A big problem in statistical causal-inference is that randomization may not be possible for physical or ethical reasons. A case study follows.

EXAMPLE 6   Since about 1920 there has been a dramatic increase in the rate of lung cancer in men. This rise also corresponds to a great increase in cigarette smoking in men. But can we conclude that cigarette smoking is a cause of the increase in lung cancer?

To hard-pressed statisticians, randomization is not feasible for this instance, and confounding factors (such as genetic defects) are always possible. Here is an example (FPP, 1978, p. 10–11): "Does smoking cause cancer? Is smoking the guilty factor? Smoking probably isn't good for you, but the verdict must be: not proved." A very fine mathematician was anguished at this verdict. "Given all the evidence how can FPP say this?" The mathematician worried that FPP's verdict might spoil the young people and concluded: "One of these three authors must be a smoker!"

Among certain statisticians, the issue of smoking and lung cancer was bitter and unsettling. Its history can be traced back to R. A. Fisher's (1959) book, *Smoking. The Cancer Controversy*. Tons of "evidence" were presented but dismissed by hard-line statisticians (see, e.g., Eysenck, 1980) on the ground that those studies were based on statistical correlation, not controlled experiments.

For instance, some researchers pointed out that the mortality ratio of lung cancer for smokers vs. nonsmokers is 10.8, seemingly strong evidence that smoking does cause lung cancer. But other researchers quickly found a mortality ratio of 0.66 for colorectal cancer and a ratio of 0.26 for Parkinson's disease. Another scientist also found that among 138 patients with Parkinson's disease, only 70% had ever smoked, compared with 84% of 166 nonsmokers who do not have Parkinson's disease (Eysenck, 1980, p. 17–18), a finding which seems to suggest that smoking will reduce the rate of Parkinson's disease.

For statisticians, it is established that many diseases are linked with smoking. But it is troublesome if we take it for granted that lung cancer, heart disease, and many other lethal illnesses are the direct consequence of smoking cigarettes.

In a private conversation in 1985, Freedman said that he had finally accepted that smoking causes lung cancer, because tobacco smoke condensate painted on the skin of mice produces skin cancers.[1] (See also Freedman, 1987). This led a friend of mine to comment that "after all, the conclusions in Fisher (1959) and FPP (1978) were wrong."

However, the hard-nosed stand of Fisher and FPP on the issue is admirable and is indeed out of necessity.[2] The reason is that in medical science the give-or-take of statistical causal-inference can mean a difference between life and death. Without this hard-edged stance, how are we going to characterize the science of statistical inference? Is it simply Russian roulette?

III. RUBIN'S MODEL AND CONTROLLED EXPERIMENTS

Causal statements often take, explicitly or implicitly, the form "if-then-else," a mode of reasoning similar to the deductive logic formulated by Aristotle in about 400 B.C. However, it is easy to confuse "one thing *causes* another" with "one thing *follows* another."

As an example, Aristotle asserted that cabbages produce caterpillars daily—a seemingly "logical" conclusion only to be refuted by controlled experiments carried out by Francesco Redi in 1668.

Aristotle, often ridiculed for the causal analyses in his book *Physics*, might have been saved from drawing erroneous conclusions if he had learned the following motto formulated by Rubin and Holland (see Holland, 1986):

NO CAUSATION WITHOUT MANIPULATION.

This motto echoes Fisher's stand on the issue of smoking and lung cancer.

In addition, Holland (1986) went further to conclude that "examples of the confusion between attributes and causes fill the social science literature" (p. 955), and that "the causal model literature [in social science] has not been careful in separating meaningful and meaningless causal statements and path diagrams" (p. 958).

Glymour (1986) responded to Holland's assertion by saying that "there is no need for this sort of care." Glymour's major argument is that probabilistic causality (as advocated by Holland) is "counterfactual." The argument is quite cute and has gained popularity among certain statisticians (e.g., Freedman, 1987; Hope, 1987; Holland, 1987, 1988). On the other hand, some statisticians have found it too abstract to penetrate. According to Glymour, statistical causalities are *counterfactual*, in the sense that

1. Counterfactuals can be logically false:

    If X were the case then X and not X would be the case.
    [For example, with 95% probability, event A will happen.]

2. Counterfactuals can logically entail one another:

    "If X were the case then Y would be the case" entails "if X were the case then Y or Z would be the case."
    [For example, with 95% confidence, A causes B.]

3. Counterfactuals have different entailment relations than do ordinary *material conditionals*: [emphasis supplied]

[In probabilistic causality,] "If X then Y" entails "If X and Z then Y."

But [in material conditionals,] "If X were the case then Y would be the case" does not entail "If X were the case and Z were the case then Y would be the case."

For example, "If I had struck the match just now it would have lighted" is true, but "If I had struck the match just now and there had been no oxygen in the room, it would have lighted" is false. (Glymour, 1986, p.964).

Simply put, counterfactuals are not logical; neither are probabilistic causalities. An implication of Glymour's analysis is that Holland's accusation (that statistical causality in social science often is meaningless) is itself meaningless. Glymour (1987, p. 17) further asserted that

we do not assume that causal relations can be reduced to probabilistic relations of any kind, but neither do we contradict such assumptions.

Counterfactual accounts of causality, according to Glymour (1986, p. 965), "have the advantage that they seem to make it easy to understand how we can have knowledge of causal relations, and equally, to ease our understanding of the bearing of statistics on causal inference." The disadvantage is that

they appeal to unobservables—to what would be true if..., and to what goes on in possible worlds we will never see. They, therefore, present us with a mystery as to how we can know anything about causal relations.

Glymour's analysis of counterfactuals is not intended to discredit probabilistic accounts of causality; instead, he attempts to dismiss Holland's motto and the Fisherian tradition that correlations cannot all be taken as causations.

However, Glymour's arguments for such statistical anarchy are not acceptable for the following reasons. First, his "material conditionals" belong to the category of "inductive logic," which simply is not logic (but indeed conterfactual). When it comes to physics or chemistry (e.g., lighting a match), things can be so stable that they mimic deductive logic. But no matter how many times you have failed to light a match in a room without oxygen, no logic will guarantee that you will fail again the next time.

In our physical world, stability exists, but there is no certainty. This assertion does not mean that science is impossible or that our world is in chaos. Rather, it means that a *logical* justification of scientific laws is impossible (see Chapter 2, Section II, "Hume's problem"). During the course of human history, physical laws that describe nature have proven stable and have thus provided a sound foundation for all sciences. Both "material conditionals" and "probabilistic counterfactuals" rest on this same foundation, although they may have different strengths.

My second objection to Glymour's anarchism is that his account of probability is inductive, not deductive. Here "inductive probability" means the fiducial probability calculated from a batch of encountered data, and "deductive probability" means that we can deduce the logical consequences of a statistical procedure. For example, if a box contains a red marble and a blue marble, then the probability of getting a red marble is 50%. This is deductive, not inductive.

In real-world applications, of course, things are not so clear-cut. For instance, a nationwide study shows that taking an aspirin every other day can reduce the risk of a first heart attack by 47 percent (the *New England Journal of Medicine*, January 28, 1988).

The study has been widely publicized. In hindsight, many people claim that they knew the beneficial effect a long time ago. For example, Kevin Begos (*The New York Times*, March 18, 1988) complained that in a 1966 Louisiana medical journal a study had demonstrated the effect but it "was objected that the study was not definitive."

Undoubtedly the results in both studies are all counterfactuals. But why does the scientific community favor one and discriminate against another? The reason is that the new study utilized randomization and involved 22,071 physicians. A box model containing red and blue marbles can thus be used for this study, and the justification of the model is the physical act of randomization. The prediction accuracy from the model now can be controlled, and the statistical procedure can be evaluated by *precise* mathematics.[3,4]

In short, there are two kinds of statistical causalities: physical causality based on statistical manipulation (i.e., randomization), and psychological causality based on enumeration from non-experimental data. The algebra for both causalities is the same, but one follows the methodology of Francesco Redi, the other of Aristotle.

* * * * * * * * *

In logic, a syllogism is called an "abduction" if the minor premise has no proof. In the case of Glymour's anarchism, the masquerading of statistical correlation as causation is neither deduction nor induction, but abduction (i.e., plain kidnapping).

Holland (1986) proposed Rubin's model as a way of dealing with the problem of causation. In literature, the complexity of Rubin's model keeps growing. However, one could argue that the model, if stripped down, is essentially Fisher's randomized comparison model.

Holland was aware that the topic of causal inference is messy and mischievous (p. 950–958). In addition, he formulates explicitly "the Fundamental Problem of Causal Inference," which means that (strictly speaking) we *cannot* observe the effect of a treatment on a population unit. This inability is, at bottom, Hume's problem of induction.

Nevertheless, our knowledge of nature advances generation after generation. According to Holland there are two general solutions by which we can get around "the Fundamental Problem of Causal Inference": the scientific solution and the statistical solution.

The *scientific solution* is to exploit various homogeneity or invariance assumptions.

The *statistical solution* is different and makes[5] use of the population U [and the average causal effect over U, the universe] in a typically statistical way. (p. 947) [emphasis supplied]

According to Holland (pp. 948–949), the statistical solution (or Rubin's model) includes the following special cases: (1) unit homogeneity, (2) constant effect, (3) temporal stability, and (4) causal transience. Apparently this framework *includes* "the scientific solution" as a special case.

But this conclusion is misleading. In practice, scientists explore causal patterns in nature. They also use established theories and mathematical tools (e.g., differential equations) to expand the territory of human knowledge. It is beyond me how Rubin's model can accommodate the rich techniques used by general scientists. In particular, I don't know how Rubin's model can help scientists to decipher a genetic code, to develop the mathematical equation for a chemical reaction, or to pin down the cause of sunspot activities.

In statistics, the purpose of randomization is to achieve homogeneity in the sample units. Further, randomization itself is independent of any statistical formula. For these reasons, the physical act of randomization, in our opinion, should be classified as a "scientific solution," rather than a "statistical solution."

More importantly, it should be spelled out that stability and homogeneity are the foundations of the statistical solution, *not* the other way around.[6,7]

Here the stability assumption means that the contents in the statistical box-model are as stable as marbles, an assumption stated in elementary statistics books but seldom taken seriously by statistics users. The box models in many "statistical analyses," as we shall see in the subsequent sections, contain not *marbles*, but only cans of worms.

IV. RUBIN'S MODEL AND OBSERVATIONAL STUDIES

In order to survive, we all make causal inferences. Even animals make causal inferences; just imagine how a mouse will run when he sees a cat.

When drawing a causal inference, most of the time neither we nor the mouse have a controlled experiment to rely on. Rubin and Holland's motto *"no causation without manipulation"* then sounds impractical and in fact untrue in numerous situations.

Here are some examples: (1) A few years ago, my life turned up-side-down because of the death of a close friend—a cause I couldn't and wouldn't manipulate. (2) A smart move by a competitor causes a businessman to panic. (3)

The meltdown of the stock market causes severe financial losses to some of my acquaintances.

Examples like these are plentiful, and the causal links are undeniable. But how far we can generalize and apply these facts to other parts of life is a totally different story. The generalization issues can easily take us into Hume's problem of induction.

Rubin and Holland are aware of the limitation of their motto. In an active attempt to cover both experiments and observational studies under the single umbrella of Rubin's model, they stress that the key element of their motto is "the *potential* (regardless of whether it can be achieved in practice or not) for exposing or not exposing each [population] unit to the action of a cause" (Holland, p. 936). It appears that this interpretation might open the door to numerous as-if experiments in social science studies. If so, then anyone could draw illegitimate conclusions without guilt, and the motto might become much ado about nothing.

Fortunately, Rubin and Holland are statisticians who wouldn't give up the Fisherian tradition easily. In their attempt to enlarge Fisher's method to cover observational studies, Rubin and Holland (and Rosenbaum, 1984, 1988; Rosenbaum and Rubin, 1984a, 1984b) have tried very hard to maintain the rigor of their model.

For instance, Rubin and Holland insist that a characteristic that cannot be potentially manipulated is not a cause. Under this restriction, race and gender , for example, cannot be considered as causes (Holland, p. 946). Tapping the same theme, Rosenbaum and Rubin (1984a, p. 26) insisted that

> *age* and *sex* do not have causal effects because it is not meaningful to discuss the consequences of altering these unalterable characteristics. [emphasis supplied]

It is interesting to see the reaction if you ask someone whether race, sex, and age are causes of certain events. Prepare yourself. The answers may not be trivial at all.

Among academicians too, opinions on race, sex, and age as causes are unsettling. Here are some examples. Pratt and Schlaifer (1984) maintained that sex and age can have a causal effect and that the effect of sex (in some cases) can be consistently estimated because "sex is randomized" (p. 31).

Freedman, Pisani, and Purves (1978) accepted the notions of "the effect of race" (p. 16), "the effect of age" (p. 39), and "the psychological effect" (p. 8). In a later work, Freedman (1988) again uses language like "aging affects metabolic processes."

Clark Glymour (1986) expressed that he was not convinced that race and gender should be excluded from the category of causes. He cited an example: "My parents told me that if I had been a girl, I would have been named Olga, instead of Clark."

Here are some other related observations: (1) In many institutions, a person will be forced to retire when he is 65 years old. (2) In other places, the message is quite clear: "People under 18 do not enter." (3) Auto insurance companies also demand higher premiums from male drivers if they are in a certain age group.

In the above examples the logical links are crystal clear. But can those examples establish age and sex to be causes? Not really. First, as scientists we are talking about natural laws, not human laws. The difference is that natural laws describe "the beings" of our universe, but human laws are the products of human decision.

Consider the example of retirement. Why do people have to retire at age 65? One reason is that the institution wants to fill the vacancies with a younger work force. But since younger people do not have working experience, what is so good about them? Also, why can we not keep both? The answer may be that the institution has budget constraints. But if money is the issue, other solutions are possible. Why must they sacrifice a valuable human resource whose experience is irreplaceable?

Questions like these can go on forever. But certainly the notion that age is *the* cause of forced retirement cannot sustain any challenge.

Another line of argument about age is more sophisticated. The argument invokes the Second Law of Thermodynamics which implies that our universe and any isolated system in it will decay toward a state of increasing disorder. A "logical" consequence of the Second Law is: Aging is inevitable, and death is certain.

But *is* death certain? Not quite. In the case of a single-cell animal, multiplication by fission is not death. And for more complex life forms, if we can keep replacing sick parts, an animal is not bound to die, and the so-called "aging" may never happen.[8]

•   •   •   •   •   •   •   •   •

It should be noted that the causal mechanisms underlying an experiment or an observational study are essentially *the same*. In other words, the difference between experiment and observational study is largely a question of states of knowledge.

This fact was promoted by Rubin, Holland, and Rosenbaum. Their stand on the issue is summarized in Holland (1986, p. 970):

> there is no difference in the conceptual framework that underlies both experiments and observational studies—Rubin's model is such a framework.

The above announcement can mean a very important scientific breakthrough. But how does the trick work? The key is a term called "strong ignorability" in Holland (p. 949) and Rosenbaum (1984, etc.).

According to Rosenbaum (1984, p. 42), a treatment is strongly ignorable if
(1) the responses, r(o), r(1), . . . , r(T), are conditionally independent of the
treatment assignment z given the observed covariates X, and (2) at each value
of X, there is a positive probability of receiving each treatment.

In mathematical notations, conditions (1) and (2) can be formulated as follows:

$$Pr[z|X, r(o), r(1), \ldots, r(t)] = Pr[z|X],$$

$$0 < Pr[z = t|X] < 1, \text{ for } t = 0, \ldots, T, \text{ for all } X.$$

In plain English, "strong ignorability" means that the way treatments were
assigned to units can be ignored in statistical computation (if X is given). For
example, in a randomized comparison, X denotes the codes of different populations, and causal inference can be drawn straightforwardly by the use of statistical formulas.

In observational studies, the process by which treatments were assigned to
units often is poorly understood, and thus "strong ignorability" may not apply.

It would be wonderful if "strong ignorability" could be *tested* by certain
statistical formulas. Regarding this important task, Rosenbaum (1984, pp. 45)
wrote:

> If the causal mechanism is correct and, in particular, if
>
> f(j) = f(k) for some j, k
>
> where f(j) is a function of (r(j), X), then rejection of the hypothesis of strong ignorability given X is also rejection of the hypothesis that the unobserved covariates U have the same distribution in the various treatment groups.

After all these intellectual investments in Rubin's model (or the Rubin-
Holland-Rosenbaum model), we deserve some reward.

EXAMPLE   (Rosenbaum, 1984, p. 45–46). This example concerns the effect
of fallout from nuclear testing on childhood leukemia. It is a follow-up of a
study by Lyon et al. (1979).

From 1951 to 1958, at least 97 above-ground atomic devices were
detonated in the Nevada desert. Data on mortality rates in high-fallout counties
are shown as follows (n = the number of incidents, r = mortality rate per
100,000 person-year):

Low-exposure 1944–1950	High-exposure 1951–1958	Low-exposure 1959–1975
n = 7	32	10
r = 2.1	4.4	2.2

The data show that the mortality rate doubled from 2.1 to 4.4 and later set back to 2.2. These results seems to indicate that the nuclear fallout caused about 16 children to die of leukemia (other cancers aside). This conclusion is compatible with laboratory experiments, in which the carcinogenic effects of radiation are well established.

However, the issue becomes puzzling if we also look at the mortality rates of low-fallout counties.

	Low-exposure 1944–1950	High-exposure 1951–1958	Low-exposure 1959–1975
Low-fallout counties	r = 4.0	3.9	3.6
High-fallout counties	r = 2.1	4.4	2.2

From this table, the previous conclusion that nuclear fallout has an effect on leukemia in high-fallout areas now appears illusory. And it becomes even more puzzling if we compare the mortality rates of other cancers:

	Low-exposure 1944–1950	High-exposure 1951–1958	Low-exposure 1959–1975
Low-fallout counties	r = 4.6	4.2	3.4
High-fallout counties	r = 6.4	2.9	3.3

Taken at face value, the data in the above table suggest that the mortality rate was lowest (at 2.9) in high-fallout counties during the high-exposure period!

Lyon et al. (1979) offered several possible explanations for these numbers.

1. Migration: The residents of the high-fallout area might have migrated to the low-fallout area.
2. Wind blow: Various weather conditions might have altered the fallout to unmonitored areas.
3. Bad luck: A chance clustering of leukemia (or cancer) deaths might have occurred in a short time.

The investigators (Lyon et al.) could not control these factors, nor did they have data related to these issues.

Without collecting more data, can Rubin's model shed new light on the problem? Rosenbaum (1984, p. 41) stated:

If treatment assignment is strongly ignorable, then adjustment for observed covariates is *sufficient* to produce consistent estimates of treatment effects in observational studies. A general approach to *testing* this critical assumption is developed and *applied* to a study of the effects of nuclear fallout on the risk of childhood leukemia. [emphasis supplied]

This statement is quite a promise. And statistical methods may score another triumph in a very difficult real-life application.

In order to understand thoroughly Rosenbaum's new analysis of a batch of old data, one has to be familiar with the notation of strong ignorability. In addition, one has to understand the Poisson kernel of the log likelihood of the raw counts. Also, the reader has to follow through a "log linear model on a flat" (p. 46) for certain counts:

$$\log(m_{eyc}) = \log(N_{ey}) + u + u_{C(c)}$$

The likelihood ratio chi-square statistic for testing the fit of the model is 13.6 for 6 degrees of freedom, with a significance level of .03. To obtain a better fit, Rosenbaum considered another log linear model:

$$\log(m_{eyc}) = \log(N_{ey}) + u + u_{Y(y)} + u_{E(e)} + u_{C(c)} + u_{EC(ec)}$$

This model provided a satisfactory fit, with a significance level of .50. Finally Rosenbaum reached the following conclusions (p. 46):

1. there have been temporal changes in reported cancer mortality aside from any effects of fallout, and
2. the high and low exposure counties had different mortality rates both before and after the period of above-ground testing and, moreover, these differences followed different patterns for leukemia and other cancers.

Well, after all the hard work, the results are not much different from common-sense conclusions one might draw by simply looking at the tables. In other words, it is hard to see the logical necessity of using Rubin's model to arrive at these results (and, in my judgment, there is none).

In summary, in the attempt to derive causation from observational studies, the so-called Rubin's model (a new rising star in the statistical literature) is still in its infant stage, and it is highly probable that the model will remain so forever.

V. CAUSAL INFERENCE IN SAMPLE SURVEY AND OTHER OBSERVATIONAL STUDIES

In this section we will take up the issue of causal inference in sample surveys. Specifically, we will discuss the problem of using sampling techniques to justify a causal link in observational data. To begin with, let's examine a case history from Freedman, Pisani, and Purves (1978, pp. 38–41).

A drug study was funded by NIH (the National Institute of Health) to assess the side effects of oral contraceptives, "the pill." One issue considered by the study was the effect of the pill on blood pressure. About 17,500 women aged 17 to 58 in the study were classified as "users" and "nonusers."

Note that randomization was impossible in this case, and a direct comparison of the blood pressures of users and nonusers can be tricky. For instance, about 3,500 women had to be excluded from the comparison, because they were pregnant, post-partum, or taking hormonal medication other than the pill. This action of throwing away data appears to go against the standard teaching of "the Law of Large Numbers," but it brings us closer to the focus of the problem.

Other controls for confounding factors are needed. For instance, about 70% of the nonusers were older than thirty while only 48% of the users were over thirty. "The effect of age is confounded with the effect of the pill," FPP wrote. "And the two work in opposite directions."

As a result, it is necessary to make a separate comparison for each age group. The following table exhibits the percentage of women with systolic blood pressure higher than 140 millimeters.

	Age			
	17–24	25–34	35–44	45–58
Nonusers	5%	5%	9%	19%
Users	7%	8%	17%	30%

The interpretations offered for the above percentages is as follows (FPP, p. 40):

> To see *the effect of the pill* on the blood pressures of women aged 17 to 24, it is a matter of looking at the percents in the columns for users and nonusers in the age group 17 to 24. To see *the effect of age*, look first at the nonusers column in each age group and see how the percents shift toward the high blood pressures as age goes up. Then do the same thing for the users. [emphasis supplied]

The "effects" can also be summarized in Figure 4. The chart indicates that age and blood pressure are strongly correlated. But before we wrap up the whole story, note that the chart is similar to Fig. 5 which was used (see Section II, Example 4) by Fong and other scientists to establish "the effect of statistical training" in everyday reasoning. Note that in Section II we have refuted

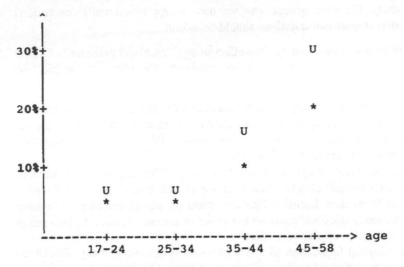

Figure 4 The effect of age on blood pressure. U: Users; *: Nonusers.

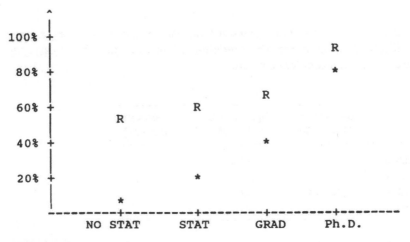

Figure 5 Percent statistical response classified by the levels of subjects' statistical training. R: Randomness cue; *: No randomness cue.

Fong's study. On what ground can we now accept Freedman's conclusion? Specifically, at least two questions should be asked:

1. What does one mean by "the effect of age" on blood pressure?
2. Does the study under discussion establish "the effect of the pill" on blood pressure?

For instance, in a laboratory we can manipulate the amount of oral contraceptives, but how are we going to manipulate different levels of age? Under the Rubin-Holland-Rosenbaum doctrine (see Section IV), expressions like "the effect of age" are smoke in logical eyes.

The second question appears easier to answer. "The physiological mechanism by which the pill affects blood pressure is well established," FPP wrote. "The Drug Study data discussed here document the size of the effect." In other words, this study does *not* establish the effect of the pill; rather, it gives rough estimate of the size of a known effect.

But the logical foundation of this conclusion is still very shaky. Recall the study about the effect of nuclear fallout on childhood leukemia (Section IV). In one case, scientists obtained the following result (r = mortality rate per 100,000 person-year):

Low-exposure	High-exposure	Low-exposure
1944–1950	1951–1958	1959–1975
$r = 2.1$	4.4	2.2

This table is compatible with laboratory experiments, in which the carcinogenic effects of radiation are *well established*. However, in another case, the scientists obtained a conflicting result:

	Low-exposure 1944–1950	High-exposure 1951–1958	Low-exposure 1959–1975
Low-fallout counties	$r = 4.6$	4.2	3.4
High-fallout counties	$r = 6.4$	2.9	3.3

which suggests that the mortality rate was lowest (at 2.9) in high-fallout counties during the high-exposure period.

As we can see now, Freedman's conclusion is not based on logic. Otherwise, the same logic would apply to the effect of nuclear fallout.

Nevertheless, we must concede that Freedman's conclusion is valid. The reasons are that (1) it is based on a large study that involved 14,000 women, and (2) the conclusion has not been challenged by other research findings.

The large sample in the drug study makes the result more convincing. But more importantly, the result is consistent with existing knowledge. By contrast, in the nuclear-fallout example, earlier results conflict with later results. Earth, as one may claim, is a very dirty laboratory, at least dirtier than human bodies.

In brief, this is how knowledge in observational studies progresses: a causal statement (such as the effect of age) is formulated to describe a general characteristic that is consistent with existing knowledge and is deemed useful for other applications. Still, the causal statement is provisional and is subject to further *refutation*. If other research findings conflict with the statement, then the statement has to be reexamined or discarded. Otherwise, the statement will be regarded as valid and useful.

This process of refutation is, in our opinion, the only way to justify a causal statement in observational study. Rubin's model or any other exotic statistical schemes provide, in most cases, not help but confusion. Cross-tabulation, regression models, and other statistical techniques are useful for the *exploration* of causal relationships. But they too do not provide justification of a causal statement.

Note that randomization is frequently used in sample surveys. But its function has nothing to do with causal inference. Rather, it is for the generalization from the sample to the target population. For this reason, causal statements in a randomized sample survey have to be classified as a subcategory of observational study.

•   •   •   •   •   •   •   •   •

In scientific inquiry, controlled experiments usually produce more reliable causal conclusions than observational studies. But this is not a rule. There are numerous exceptions.

For example, in astronomy we cannot manipulate any quantities whatever, yet predictions in astronomy are often more accurate than those produced by controlled experiments in clinical trials.

In a personal conversation at the 1987 ASA Annual Meeting, Holland stunned me by saying that modern astronomers usually carry out a lot of experiments before they draw conclusions. This keen observation unfortunately does not apply to geology. For example, drilling for petroleum is an enormous gamble. When a geological study indicates the place where petroleum might have accumulated, there is less than a 10% chance that oil is actually present; and there is *only a 2% chance* that it is present in commercially useful amounts. (Professional drillers sometimes joke that "random drilling may be better.")

The intrinsic unpredictability of many observational studies is beyond any logic. The gap between prospective and retrospective analyses is also unbridgeable. We therefore disagree with Holland's contention that Rubin's model (or anybody's model) is a framework for both experiments and observational studies. Holland has also asserted that "the idea that Rubin's model is somewhat incapable of accommodating time-series data is misleading." In my opinion, it is Holland's statement that is misleading. For example, if Holland hasn't forgotten his own motto, I hope he can tell us how he can literally manipulate time-series data or any encountered data in geoscience.

For randomized experiments, causal inferences can be drawn straightforwardly by statistical bulldozer. For observational studies, causal inferences have to be assessed on a case-by-case basis; a generic approach like Rubin's model may create more confusion than the clarity the model originally intended.

For those who are serious about identifying the cause from encountered data, a piece of advice from R. A. Fisher (quoted from Rosenbaum, 1984, p. 43) usually proves helpful: *Make your theories elaborate*.

This advice may be foreign to some statistics users. However, the practice of the maxim is an integral part of process quality control, an orphan in traditional statistical training.

For instance, when problems emerge in a manufacturing process, how do engineers set forth to find the cause and solve the problem? They explore related scientific knowledge. More specifically, the techniques include brainstorming, Ishikawa charts, Pareto diagrams, and statistical control charts.

Here is a beautiful example from Grant and Leavenworth (1980, p. 95). In this case study, statistical control charts were maintained to monitor the quality of specially treated steel castings. For a number of months, the final products of castings did meet rigid requirements. But all of a sudden points on $\bar{x}$-charts went out of the control limits.

The supervisors in the production department were sure no changes had been made in production methods and at one point attempted to blame on the testing laboratory. Unfortunately, independent tests showed the points continuing to fall out of the control limits.

Pressure mounted on production personnel, and they were forced to explore potential causes further. This activity of finding the cause or causes usually is very frustrating; but for competent scientists, it can also be as challenging (and rewarding) as Sherlock Holmes' adventures.

To conclude this case study, finally somebody noted that a change in the source of cooling water for heat treatment had been made just before the time the points fell out of control charts. Nobody believed that a change of cooling water could have created so much trouble. However, as a last resort, the original source of quench water was restored. Immediately the points fell within the control limits, and the process remained under control at the original level.[9] To sum up the whole story, the great detective had it all (Sherlock Holmes, *The Sign of the Four*):

> How often have I said to you that when you have eliminated the impossible, whatever remains, however improbable, must be the truth.

VI. CAUSES, INDICATORS, AND LATENT VARIABLES

In social or behavioral sciences, unmeasured latent variables often are created to account for a large portion of variability in the data. The statistical techniques used to construct these latent variables are primarily principal components and factor analysis (see, e.g., Morrison, 1976). In the case of factor analysis, the trick goes like this. Let X_1, X_2, . . . , X_p be *observable* random variables; next let Y_1, Y_2, . . . , Y_m ($1 \leqslant m < p$) be unmeasured latent variables that will be used to construct a simpler structure to account for the correlations among the Xs:

$$X_1 = a_{11} * Y_1 + \cdots + a_{1m} * Y_m + e_1$$

$$\cdots$$

$$X_p = a_{p1} * Y_m + \cdots + a_{pm} * Y_m + e_p$$

Occasionally one is lucky: $m = 1$. Some examples of latent variables are the so-called "cognitive ability" or "socioeconomic status"; the latter is routinely used in sociological studies as a latent variable that affects education, income, occupation, and other observable variables (Glymour, 1987).

Latent variables of this sort appear to be outside the sphere of Rubin's model. In numerous cases, these variables are useful constructs for policy- or decision-making. But a nasty question has been haunting the research workers in the field: Can the latent variables be considered "causes?"

To answer this question, some examples from natural science are worth retelling. To begin with, let's consider the story of John Dalton, an Englishman who has been widely recognized as the father of modern chemistry (for his contribution to the theory of the atom).

Greek philosophers (such as Democritus) are sometimes credited for their invention of the notion of "atom." But Dalton's contribution (1803) was different, because it was based on the laws of conservation of mass and definite proportions—laws that had been derived from direct observation and had been tested rigorously. The theory can be expressed by the following postulates (Brady and Humiston, 1982):

- All elements are made up of atoms which cannot be destroyed, split, or created.

- All atoms of the same element are alike and differ from atoms of any other element.
- Different atoms combine in definite ratios to form molecules.
- A chemical reaction merely consists of a reshuffling of atoms from one set of combinations to another. The individual atoms remain, however, intact.

Dalton's theory was wrong on numerous counts; but the theory proved successful in explaining the laws of chemistry and was embraced by the scientific community almost immediately. We now know that in one drop of water there are more than 100 billion billion atoms. With a quantity this small, how did Dalton manipulate an atom? Specifically, how did he gauge the weight of an atom, or measure the distance between atoms in a crystal of salt? The answer is that none of these tasks was accomplished by Dalton or his contemporaries.

As a matter of fact, the doubt of the very existence of atoms and molecules did not rest completely until the systematic measurements of the sizes of atoms were carried out in 1908 by J. B. Perrin (the winner of the 1926 Nobel Prize for Physics). This confirmation occurred 105 years after Dalton's original proposal.

Dalton's concept of atom was, in his time, an unmeasured latent factor; but it was accepted as a fruitful working hypothesis by "reasonable" scientists. Without Dalton's theory, modern chemistry and physics might have had to wait until the next century.

The second example of a latent factor is related to the discovery of an important elementary particle. In the subatomic world, protons (p) may transmute into neutrons (n), and vice versa. The reactions are called beta-decay, which involves the release of an electron (e) or antielectron ($\bar{e}$):

$$n \longrightarrow p + e$$

$$p \longrightarrow n + \bar{e}$$

A curious feature of beta-decay is that the emitted electrons or antielectrons sometimes emerge with one energy and sometimes with another. This phenomenon prompted some theorists to speculate on the failure of the conservation law of energy. In 1931 Wolfgang Pauli proposed that an elementary particle (now called "neutrino") is the cause of the variability of the energy of beta-rays. At the time there was no experimental evidence for the existence of any extra particle in beta-decay, and Pauli's proposal was only *an act of faith*, based on nothing but a firm belief in the conservation laws (Ohanian, 1987, p. 407). Experimental evidence for the existence of neutrinos was not available until 1953, 22 years after Pauli's original proposal.

A more spectacular example than Pauli's neutrino is Maxwell's electromagnetic waves. Prior to Maxwell's work, Faraday found that moving a magnet through a coil of copper wire caused an electric current to flow in the wire. The discovery led to the development of electric generator and motor. In a striking fashion, Maxwell used the experimental discoveries of Faraday to arrive at a set of equations depicting the interaction of electric and magnetic fields. On the basis of these equations, he predicted (in 1884) the existence of electromagnetic waves that move through space with the speed of light. This prediction was confirmed 23 years later by H. Hertz (World Book, 1982). In his experiment, Hertz used a rapidly oscillating electric spark to produce waves with ultra-high frequencies; he also showed that these waves cause similar electrical oscillations in a distant wire loop. The manipulation of these waves later led to the invention of radio, television, and radar.

In the above examples, the latent factors could not be directly measured or manipulated when they were initially introduced to modern science. For this reason, Rubin and Holland's motto, "no causation without manipulation" may need a more flexible interpretation. In short, if a theory is sound and indirect confirmations do exist, a man-made causal schema like Dalton's atomism deserves the respect and attention of working scientists.

In social and behavioral sciences, the statistical techniques for extracting latent variables have been a constant target of criticism. One of the most famous rejections of latent variables came from B. F. Skinner (1976), the god-father of behaviorism. Simply put, Skinner maintained that anything that can be explained with latent variables can be explained without them (a la Glymour, 1987).

In opposition to Skinner's and Holland's rejections of latent variables as legitimate causes, Glymour (1987) responded that "the natural sciences are successful exactly because of their search for latent factors affecting the phenomena to be explained or predicted." To substantiate this claim, Glymour argued:

> Newtonian dynamics and celestial mechanics, the theory of electricity and magnetism, optics, chemistry, genetics, and the whole of modern physics would not have come to pass if natural scientists [had] behaved as the critics of latent variables prefer.

> Gravitational force, electrical fluids and particles, electromagnetic fields, atoms and molecules, genes and gravitational fields, none of them could be directly measured or manipulated when they initially became part of modern science.

Glymour is quite right on this mark. But one might feel uncomfortable when he extrapolates the success of latent factors in natural science to other branches of science:

> Critics may grant that the introduction of latent variables has been essential in the natural sciences, but maintain that they are inappropriate in the social and behavioral sciences. It is hard to think of any convincing reason for this view.

Glymour is convinced that heuristic search, utilizing artificial intelligence and statistical correlation matrices, could produce latent variables for important scientific discoveries (Glymour, 1987, p. 23):

> In Part II of this book we will show that linear causal models can explain *in the same fashion* that, say, Daltonian atomism explained the law of definite propotions, that Maxwell explained electromagnetic phenomena, or that Copernican theory explained regularities of planetary motion. [emphasis supplied]

Scientific laws, as it appears, can now be easily cloned by statistical correlation matrices. Figure 6 is a typical example of Glymour's search for a causal structure about industrial and political development (p. 148). GNP is gross national product, Energy is the logarithm of total energy consumption in megawatt hours per capita, Labor is a labor force diversification index, Exec is an index of executive functioning, Party is an index of political party organization, Power is an index of power diversification, CI is the Cutright Index of political representation, and the e's are the added error terms.

To shorten the notations, let ID = industrial development and PD = political development. Then the path from ID to GNP and the path from ID to Energy, for example, are equivalent to the linear statistical equations:

$$GNP = a_1 * ID + e_1$$
$$Energy = a_2 * ID + e_2$$

A revised model (Fig. 7) can be achieved by adding more paths to the original skeleton. The path regression for ID- > GNP and ID- > Exec read as follows:

$$GNP = a_1 * ID + e_1$$
$$Exec = a_4 * ID + b_4 * PD + e_4$$

Note that in this model, ID and PD are latent variables, while the rest are measured quantities. The last equation is thus a linear statistical law relating a measured variable with two unmeasured variables.

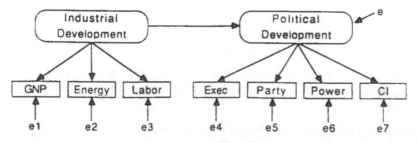

Figure 6   Graph representation for industrial and political development. Copyright © 1987 by Academic Press, Inc. Reprinted by permission.

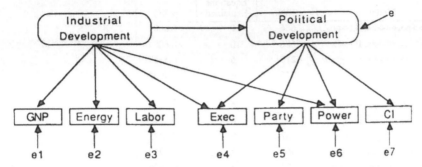

Figure 7 Revised model for industrial and political development. Copyright © 1987 by Academic Press, Inc. Reprinted by permission.

The fit of the revised model is marginal, with a P-value = .053 for a chi-square test. Naturally one would look for alternatives or to improve the model by adding further connections. With the aid of a computer program called TETRAD (Glymour, p. 162), 38 models were located and ordered by the probability of their chi-square statistic see Fig. 8). In this figure, TTR is a TETRAD measure of the goodness-of-fit, which, according to Glymour, is negatively correlated with the chi-square test. But one can easily detect the inconsistency of these measures in the pairs of models such as (M2,M3), (M4,M5), (M3,M37), etc.; where M2 stands for model 2.

In one of his concluding remarks, Glymour wrote,

> To justify the assumptions of the [best] model, we need either substantive knowledge or grounds for thinking that no other plausible model explains the data as well.

This conclusion certainly should be underlined many times by all statistics users. But the bottom line is: "What is the causal structure of all these variables?" In Glymour's discussions of the computer outputs, we find nothing clear except the following: (1) the substantive knowledge was weak, (2) the measurements of the variables (such as Exec, an index of executive functioning) and the measures of goodness-of-fit are rough, (3) one may find infinitely many other plausible models by adding more variables, more connections, time-lag effect, and more sophisticated transformations.

In sum, one can hardly see any causal structure from these models. And one will be better off if one does not take these models seriously as scientific laws. To support this position, in Chapter 4, under the title of "Amoeba Regression," we will present further examples to demonstrate the structureless nature of such statistical models.

Similar observations about path analysis were made in Freedman (1985, 1987). After careful and interesting elaborations, Freedman rejected flatly the idea that linear statistical laws in social science are at all comparable to

All Models Located with TETRAD			
Model	TTR	Equations Explained	$p(X^2)$
1) en->po, en->ex, ID->po	.686	10	0.6082
2) en->po, en->ex, ID->ex	.686	10	0.6004
3) en->po, en->ex, gp->pa	.270	5	0.5752
4) en->po, en->ex, pa->en	.246	4	0.5680
5) en->po, en->ex	.686	10	0.5504
6) en->po, en->ex, ci->la	.251	4	0.5198
7) en->po, en->ex, la->pa	.202	6	0.4792
8) en->po, la->pa, po->ex	.202	6	0.4085
9) ID->po, ID->ex, PD->en	.560	5	0.2940
10) ID->po, en->ex, pa->en	.328	5	0.1526
11) ex->en, en->po			0.1434
12) en->po, la->ex, pa->en	.213	3	0.1234
13) ID->po, en->ex, ex->gp	.358	5	0.1222
14) en->po, gp->ex, en->pa	.220	6	0.1189
15) en->po, la->ex, ex->pa	.223	5	0.1108
16) en->po, gp->ex, po->pa	.220	6	0.1002
17) ID->po, po->en, ex->en	1.670	8	0.0931
18) en->po, la->ex, la->pa	.222	5	0.0819
19) en->po, gp->ex, la->pa	.146	7	0.0551
20) ID->ex, ID->po	1.149	11	0.0530
21) ID->ex, po->en, ex->en	1.364	8	0.0477
22) en->ex, pa->po, la->pa	.272	5	0.0438
23) en->ex, po->en			0.0336
24) en->ex, la->po, pa->la	.198	4	0.0246
25) en->ex, pa->po, en->pa	.286	6	0.0211
26) en->ex, la->po, ci->la	.198	4	0.0161
27) T->ex, T->en, T->po, T->ID, T->PD	.565	8	0.0145
28) en->ex, gp->po, pa->en	.237	4	0.0135
29) en->ex, gp->po, la->pa	.206	5	0.0133
30) en->ex, gp->po, ci->en	.237	4	0.0108
31) en->ex, pa->po, ex->pa	.286	6	0.0096
32) T->ex, T->en, T->po	2.345	7	0.0015
33) ex C en, en C po, ex C po	2.345	7	0.0015
34) gp->ex, pa->po, en->pa	.220	6	<0.001
35) ex->en, po->en			<0.001
36) gp->ex, gp->po, la->pa	.220	6	<0.001
37) gp->ex, gp->po, la->pa	.209	6	<0.001
38) Skeleton	5.300	13	0.000

Figure 8 All models located with TETRAD. Copyright © 1987 by Academic Press, Inc. Reprinted by permission.

Maxwell's equations or Copernicus-Kepler astronomy. One reason is that the stochastic assumptions embodied in path regression do not stand on water. Another reason is that few investigators in social science do careful measurement work. "Instead, they factor-analyze questionnaires. If the scale is not reliable enough, they just add a few more items" (Freedman, 1985).

•　　•　　•　　•　　•　　•　　•　　•　　•

In many observational studies, causal inferences are plain speculations. In the history of science, such speculations have often turned out to be the seeds of important discoveries. For this reason, we believe it is worthwhile to provide another motto:

NO CAUSATION WITHOUT SPECULATION

This motto is a complement, not a rival, to Holland's motto. In short, speculation breeds scientific theory, but the theory has to be *tested* rigorously against reality. As any insider can testify, in the history of science, many beautiful theories had been put together, only to be shot to pieces by a conflict with experiment (see Will, 1986). Experimentation is often called "the Queen of Science." Certainly no exaggeration.

But a word of caution is that statistics users often equate the word "test" with statistical testing. As an example, Granger (a towering figure in econometric research) wrote (1986): "If causality can be equated with some measurable quantity, then *statisticians should be able to devise tests for causation*" [emphasis supplied].

As a statistician, I feel honored that statistics users have this kind of confidence in my profession. On the other hand, I am also deeply troubled, because in most cases a scientific theory of cause-and-effect has to be tested by the scientists themselves, not by statisticians. Statisticians have indeed made important contributions to numerous instances of scientific testing; but their role is often blown out of proportion.

Consider this example about the effect of gravitational force on a falling object (*Scientific American*, March 1988). Newton and Einstein both maintained that an object's gravitational acceleration is independent of its mass and substance. New theories challenge this fundamental notion. But how can we put the issue to a test? Can we do so by calling in a statistician to perform a hypothesis testing? Not a chance.

Not only in physics and chemistry, but also in social science studies, statistical methods alone yield no golden promise as Granger hopes. As an example, sending people to school is often believed to increase their income. But how can we measure the effect and "devise tests for causation" (a la Granger)? The solution is not straightforward, if it exists at all.[10]

NOTES

1. Freedman, Pisani, Purves, and Adhikari (1991, p. 12, *Statistics*, Norton) concluded that smoking causes heart attacks, lung cancer, and other diseases.
2. In fact, a recent study (*Science News*, June 4, 1988) provides evidence that an absence of tumor-suppressor genes may lead to lung cancer.
3. Another study involving 5,000 British physicians failed to find any similar benefits from daily aspirin intake (*Science News*, February 6, 1988). In comparison, the U.S. study involved more than 22,000 test subjects and three times as many heart attacks. The law of large numbers gives the U.S. study a winning edge.
4. The study does not provide information on the side effects of aspirin. For this purpose, one needs other studies.
5. This definition of "the statistical solution" to the problem of causal inference is correct. But one has to be aware that "the universe" of a typically randomized study consists only a handful of subjects.
6. For instance, in a clinical trial, applications of a randomized study to new patients rely on both the stability and homogeneity assumptions of our biological systems.
7. One may argue that the stability assumption can be (and should be) tested by statistical methods. This is true in cases that involve reliability theory and life testing. In most applications, stability assumption is established by enumeration, and does not involve any formal statistical inference.
8. The issues involving race and sex are more controversial, because in some cases the notions that race and sex are causes do have a certain genetic basis.
9. Ingenious applications of this kind are plentiful. See, for example, Ott (1978), *Process Quality Control, Trouble-shooting and Interpretation of Data*, pp. 2–6, 34–37, 57–58, 80–82, 110, 112–115, 116–118, 130–132, 146, and 171–173.
10. See Freedman (1987, pp. 103–104).

REFERENCES

Baron, J. and Hershey, J. C. (1988). Outcome Bias in Decision Evaluation. *J. of Personality and Social Psychology*, Vol. 54, No. 4, 569–579.

Brady, J. E. and Humiston, G. E. (1982). *General Chemistry: Principle and Structure*, 3rd edition. Wiley, New York.

Eysenck, H. J. (1980). The Causes and Effects of Smoking. Sage, Bervely Hills, California.

Fisher R. A. (1959). *Smoking. The Cancer Controversy*. Oliver and Boyd, Edinburgh.

Fong, G. T., Krantz, D. H., and Nisbett, R. E. (1986). The Effect of Statistical Training on Thinking about Everyday Problems. *Cognitive Psychology*, Vol. 18, 253–292.

Freedman, D. A. (1985). Statistics and the Scientific Method. *Cohort Analysis in Social Research*, edited by W. M. Mason and S. E. Fienberg. Springer-Verlag, New York, pp. 343–390..

Freedman, D. A. (1987). As Others See Us: A Case Study in Path Analysis. *J. of Educational Statistics*, Vol. 12, No. 2, 101–128, 206–223 (with commentaries by 11 scholars, 129–205).

Freedman, D. A., Pisani, R., and Purves, R. (1978/1980). *Statistics*. Norton, New York.

Freedman, D. A. and Zeisel, H. (1988). Cancer Risk Assessment: From Mouse to Man. *Statistical Science*, Vol. 3, No. 1, 3–28, 45–56, (with commentaries, 28–44).

Glymour, C. (1986). Statistics and Metaphysics. *JASA*, Vol. 81, No. 396, 964–966.

Glymour, C., Scheines, R., Spirtes, P., and Kelly, K. (1987). *Discovering Causal Structure: Artificial Intelligence, Philosophy of Science, and Statistical Modeling*. Academic Press, New York.

Granger, C. (1986). Comment on "Statistics and Causal Inference," by P. W. Holland. *JASA*, Vol. 81, No. 396, 967–968.

Grant, E. L. and Leavenworth, R. S. (1980). *Statistical Quality Control*. Fifth Edition. McGraw-Hill, New York.

Holland, P. W. (1986). Statistics and Causal Inference, *JASA*, Vol. 81, No. 396, 945–960 (with commentaries by D.B. Rubin, D.R. Cox, C. Glymour, and C. Granger, 961–970).

Holland, P. W. (1988). Causal Inference in Salary Discrimination Studies. Presented at the 1988 ASA Annual Meeting, New Orleans.

Holland, J. H., Holyoak, K. J., Nisbett, R. E., and Thagard, P. R. (1987). *Induction*. MIT Press, Cambridge, Massachusetts.

Hope, K. (1987). Barren Theory or Petty Craft? A Response to Professor Freedman. *J. of Educational Statistics*, Vol. 12, No. 2, 129–147.

Lyon, J. L., Klauber, M. R., Gardner, J. W., and Udall, K. S. (1979). Childhood Leukemias Associated with Fallout from Nuclear Testing. *New England Journal of Medicine*, Vol. 300, No.8, 397–402.

Moore, D. (1988). Comment on "The Teaching of Statistics," by H. Hotelling. *Statistical Science*, Vol. 3, No. 1, 84–87.

Morrison, D. F. (1976). *Multivariate Statistical Methods*. McGraw-Hill, New York.

Ohanian, H. C. (1987). *Modern Physics*. Prentice Hall, Englewood Cliffs, New Jersey.

Ott, E. R. (1975). *Process Quality Control: Trouble-Shooting and Interpretation of Data*. McGraw-Hill, New York.

Pratt, J. W. and Schlaifer, R. (1984). On the Nature and Discovery of Structure. *JASA*, Vol. 79, No. 385, 9–21, 29–33, (with commentaries, 22–28).

Rosenbaum, P. R. (1984). From Association to Causation in Observational Studies: the Role of Tests of Strongly Ignorable Treatment Assignment, *JASA*, Vol. 79, 41–48.

Rosenbaum, P. R. (1988). Detecting and Assessing Bias in Observational Studies. Presented at the 1988 ASA Annual Meeting, New Orleans.

Rosenbaum, P. R. and Rubin, D. B. (1984a). Estimating the Effects Caused by Treatments. *JASA*, Vol. 79, No. 385, 26–28.

Rosenbaum, P. R. and Rubin, D. B. (1984b). Reducing Bias in Observational Studies Using Subclassification on the Propensity Score. *JASA*, Vol. 79, No. 387, 516–524.

Skinner, B. F. (1976). *About Behaviorism*. Vintage Books, New York.

Will, C. M. (1986). *Was Einstein Right?* Basic Books, New York.

Chapter 4

Amoeba Regression and Time-Series Models

I. DISCOVERING CAUSAL STRUCTURE: SCIENCE NOW CAN BE EASILY CLONED

In 1987, as we have mentioned in Chapter 3, Glymour and coworkers (Scheines, Spirtes, and Kelly) published a book entitled *Discovering Causal Structure: Artificial Intelligence, Philosophy of Science, and Statistical Modeling*. This ambitious work was funded by three NSF grants. A major goal of the book is to uncover causal structure. For this task, the group headed by Glymour developed a computer program called TETRAD, which is purported to generate scientific theory (p. 58) by peeking into a specific set of data where substantive knowledge of the domain is whistling in the dark (Glymour, p. 7). This attempt is admirable, but it is certainly in sharp contrast to a popular belief (Albert Einstein, 1936) that "theory cannot be fabricated out of the results of observation, but that it can only be invented."

It is well known that experiments are not always practicable (or ethical) and that even when they are, they may not answer the questions of interest. "Faced both with urgent needs for all sorts of knowledge and with stringent limitations on the scope of experimentation," Glymour (p. 3) wrote, "we resort to statistics."

With great confidence in statistical modeling and TETRAD, Glymour (p. 13) asserted: "A TETRAD user can learn more in an afternoon about the existence and properties of alternative linear models for covariance data than could an unaided researcher in months or even years of reflection and calculation."

Some scientists and statisticians oppose TETRAD or the like. Their arguments and Glymour's forceful defense (see Glymour, 1987, pp. 9, 15–59, 234–246, etc.) constitute a lively debate, and the battles of different schools of thoughts are likely to continue for years to come.

With the aid of Glymour's powerful computer package and his dazzling arguments of epistemology (the theory of knowledge), such statistical models may become a new fashion among certain researchers. But as a statistician, I have mixed feelings about Glymour's ambitious program. On the one hand, I am willing to accept the TETRAD activities under the new motto, "No causation without speculation." On the other hand, if TETRAD delivers only an empty promise, the reputation of my discipline will suffer. For this reason, I conducted an experiment to test Glymour's promise.

A set of triplets was created and given to the students in my regression class:

X_1	X_2	Y
2	2.75	3.42
2	2.50	3.21
2	2.25	3.01
2	2.00	2.89
2	1.80	2.75
2	1.50	2.55
2	1.25	2.38
2	1.00	2.30
2	0.70	2.15
2	0.50	2.10
3	2.50	3.95
3	2.00	3.63
3	1.50	3.40
3	1.00	3.18
4	1.75	4.38
4	1.50	4.29
4	1.20	4.18
4	0.80	4.10
5	1.75	5.30
5	1.50	5.22
5	1.00	5.15
5	0.75	5.08

I told the students that the background of the data would not be revealed and that they were allowed to use any method to find an equation to fit the data. A few hints were available: (1) Y is a function of X_1 and X_2. (2) There is no serial correlation among the data; thus the Durbin-Watson test and time-series modeling of residuals were not necessary.

The students were excited about the project; it was like a treasure hunt. To begin with, they all plotted the data in scatter diagrams (see Fig. 1). The Pearson correlation coefficients are .9476 for (X_1, Y) and 0.054 for (X_2, Y), with corresponding P-values of .0001 and .810, respectively. Yet a possible relationship between X_2 and Y could not be rejected out of hand, because an enhanced diagram reveals a certain pattern (see Fig. 2). The scatter diagrams indicate that a simple algebraic function might suffice. Nevertheless, my students tried different transformations of the variables. After elaborate screening procedures, each student presented his/her best models. Here are some of them (the null hypothesis for each P-value being that the coefficient is zero):

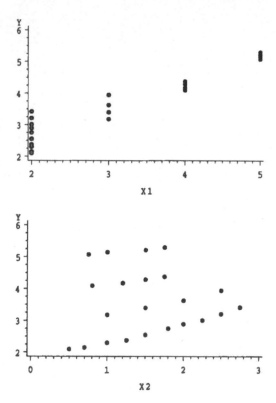

Figure 1   Scatter diagram for Y vs. X_1 and Y vs. X_2.

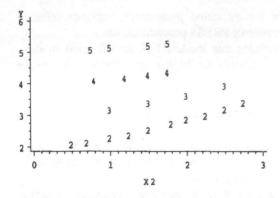

Figure 2 Enhanced scatter diagram for Y vs. X_2.

Model	R-square
1. $Y = (.199)X_1^2 + (.420)X_2^2$ P: .0001 .0001	96.36%
2. $Y = (.897)X_1 + (.525)X_2$ P: .0001 .0001	99.93%
3. $\log(Y) = (-.60) + (.9)\sqrt{X_1} + (.18)X_2$ P: .0001 .0001 .0001	98.29%
4. $\log(Y) = (.174) + (.25)X_1 + (.18)X_2$ P: .0001 .0001 .0001	98.17%
5. $\log(Y) = (.282) + (.257)X_1 + (.057)X_2$ P: .0001 .0001 .0001	98.3%
6. $Y = (.247) + (.914)X_1 + (.0897)*\exp(X_2)$ P: .0001 .0001 .0001	98.46%
7. $Y = (.678) + (.867)X_1 + (.659)*\log(X_2)$ P: .0001 .0001 .0001	98.06%
8. $\log(1/(Y-1)) = (.70) + (-.36)X_1 + (-.29)(X_2)$ P: .0001 .0001 .0001	97.08%
9. $Y = (1.03) + (.82)X_1$ P: .0001 .0001	89.8%

The students were required to perform a comprehensive diagnostic checking of their models, which includes residual plots, stem- leaf and box-whisker plots, normal probability plot, Shapiro-Wilk normality test, F-test, Cook's influence statistics, correlation matrix of the estimated parameters, variance inflation factors, cross-validation (D. M. Allen's PRESS procedure), etc.

These diagnostic statistics indicate that models 1–8 are all good models. Further, the R-squares of these models are higher than 96%—a lot better explanatory power than most models in social science studies. It appears that any of these models should perform well, at least for the purpose of prediction. "But what is the underlying mechanism that generated the data?" my students asked.

Well, a few days before the assignment, I sat at home and drew 22 right triangles like those in Fig. 3. The assignment the students received was to rediscover the Pythagorean theorem, and they all failed.[1]

This experiment led me to coin a new term "amoeba regression," meaning that the causal structures "discovered" by such regression techniques may be as structureless as an amoeba. The experiment also forced me to ponder the role of statistical modeling in social science research.

It was not an overstatement when someone said that "mathematics is the mother of science." Without Greek mathematics (especially Euclidean geometry), I cannot imagine what Newtonian mechanics would be. Without Newtonian mechanics, I cannot imagine what modern physics and technology would be. Moving back to the fundamentals, without the Pythagorean Theorem, I cannot imagine what Euclidean geometry would be.[2] In short, if a scientific discipline indulges in linear regression but lacks anything fundamental, the discipline may have serious problems.

The above experiment did not accommodate sophisticated transformations (for example, the celebrated Box-Cox transformation). Out of curiosity, I conducted another regression analysis, assuming that a fictitious investigator squares the Y-variable and then fits the second order Taylor expansion:

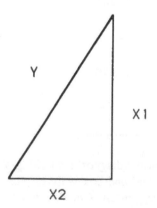

Figure 3   One of the triangles used to generate the data for (Y, X_1, X_2).

$$Y^2 = -.125 + (.12)X_1 + (.28)X_2 + (.99)X_1^2 + (.96)X_2^2 + (-.07)X_1 * X_2$$

P: .80   .67   .46   .0001   .0001   .28

Note that only the quadratic terms are statistically significant. The investigator then refits the data as follows:

$$Y^2 = (1.007)X_1^2 + (1.02488)X_2^2$$

P   <   .0001   .0001

SE: .00308   .01229

This model passes the scrutiny of diagnostic checking with an R-square of 99.99%. The model almost recovers the Pythagorean formula.[3]

But there are other profound troubles in using Taylor expansion and regression analysis to "discover" fundamental laws. First, by what criterion can we be sure that the last model is better than the others? To my knowledge, current technology provides no clear-cut solution. Second, If the legs of the triangles are not orthogonal, then

$$Y^2 = X_1^2 + X_2^2 - 2 * X_1 * X_2 * \cos \theta$$

where θ is the angle between the vectors $\vec{X}_1$ and $\vec{X}_2$. I think the chance that a shotgun regression would rediscover this formula is minuscule.

Like many statistics users, I used to believe that if a model passed a comprehensive diagnostic checking, the model should be useful, at least for the purposes of prediction. The above amoeba regressions shake this belief. The experiment indicates that regression models cannot all be taken as seriously as scientific laws. Instead, they share similarities with the "common sense" in our daily life: informative but error-prone, depending on the skills of the users. Since statistics are often called "educated guesses," I propose to call these regression models "educated common sense."

Glymour's TETRAD produces, at best, "descriptive models," which may or may not be useful, depending on the model, the data, and what happens next (Freedman, 1987, p. 220).

In brief, there are two kinds of regression models: (1) structural models that represent a deep commitment to a theory about how data were generated (Freedman, 1987); and (2) descriptive models, which are the mere refinements of common sense. Although, as a statistician, I have a high respect for educated (and uneducated) common sense, I believe that it is profoundly mistaken to put the two types of models in the same bag.

Glymour's effort to build TETRAD reminds me of a fabled Chinese, Sen Nong (fl. between 3000 and 4000 B.C.), a physician who was reported to have tried several hundred species of herbs on himself in order to find useful medicine for the Chinese people.

Before modern medication was introduced to Asia, generations of Chinese were saved (and perhaps in many cases poisoned) by Sen Nong's herbal medicines. Today some oriental medications remain valuable, and their functions are still mysterious to western scientists. However, the real revolution (and the real lifesaver) in oriental history is the adoption of western medication.

The main driving force behind modern science is the quest to understand the underlying mechanisms. In medication, this means solid knowledge of physiology, pathology, biochemistry, anatomy, etc., which were developed mainly by substantive knowledge. Regression analysis certainly helps, but it is used in highly focused subject matters, under tightly controlled experiments.

By contrast, in many branches of social, economic, and behavioral sciences, solid theory and good measurement are both lacking. Research workers then take statistics (and computers) as their last and best hope. Their desperation and methodology were shared by ancient Chinese physicians. Such methods as TETRAD are, in my opinion, legitimate and useful research tools; but their roles have been greatly exaggerated. Indeed, I would be horrified if medical researchers proposed linear statistical laws, instead of physiology and pathology, as the foundation of medical practices.

As pointed out in Glymour et al. (p. 16), some social scientists prefer to avoid causal language. This attitude is highly recommended but was unfortunately dismissed by Glymour.

II. REGRESSION AND TIME-SERIES ANALYSIS: SCIENCE OR LUNACY? (Part I)

A major benefit of scientific knowledge is prediction. Since the 1950s, regression and time-series analysis have been used for economic forecasting, which is reportedly a 100–200 million dollar business. This new industry has an impressive history, which even includes the prestige of Nobel Prizes.

In 1969 Norwegian economist Ragnar Frish (a professor in social economy *and* statistics) won the first Nobel Prize in Economic Science, partly for his contribution of introducing random disturbances to deterministic dynamic systems.

In 1980 Lawrence Klein won the Prize partly for pioneering the development and use of econometric models to forecast economic trends in the United States. The Klein-Goldberger (1955) model further gave rise to numerous econometric research efforts that rely heavily on two-stage least squares, three-stage least squares, and time-series analysis.

In 1989 Trygve Haavelmo also won the Nobel Prize "for his clarification of the probability theory foundations of econometrics and his analyses of simultaneous economic structure" (the Royal Swedish Academy of Sciences; *Trenton Times*, October 12, 1989). According to Paul Samuelson, another Nobel laureate at the Massachusetts Institute of Technology,

The whole modern age in economics has tried to use *statistics* to test out different theories. It was Haavelmo who made the great breakthrough, so we can know we're testing our particular theory and are not getting a spurious relationship. [emphasis supplied]

Today the models derived from Haavelmo's work "are widely used in predicting the course of national economies and formulating government policies" (*The New York Times*, October 12, 1989). But how good are these models? And what are other experts' opinions of these models?

In a meta-analysis, Armstrong (1984/1978) provided interesting findings related to the above questions. The study involved a survey of 21 experts from some of the leading schools in econometrics (MIT, Harvard, Wharton, Michigan State, etc.) and from well-known organizations that sell econometric forecasts.

The survey was designed to examine two popular beliefs:

1. Econometric models provide more accurate short-term forecasts than do other methods; and
2. More complex econometric methods yield more accurate forecasts than do simpler methods.

Of the 21 experts surveyed, 95% agreed with the first statement, and 72% agreed with the second.

Respondents were asked how much confidence they had in their opinion on accuracy. Confidence was rated on a scale from 1 ("no confidence") to 5 ("extremely confident"). The average response was about 4. No one rated confidence lower than 3.0 (p. 23).

The experts were also asked to rate their own professional competence. Those who rated themselves as more expert felt that econometric methods were more accurate.

To assess the reliability of the experts' opinions, Armstrong went through a search of 30 studies from research journals, and the verdict on the first belief is disheartening: "Econometric forecasts were not found to be more accurate" (1984, p. 25).

Moving on to the second belief that more complex econometric methods yield more accurate forecasts, empirical evidence gathered by Armstrong from 11 direct and 5 indirect studies contradicted the folklore. Armstrong's conclusion also strengthens an argument by Milton Friedman, another Nobel laureate (1951), that "the construction of additional models along the general lines [as Klein's model] will, in due time, be judged a failure."

Many researchers have devoted their whole lives to complex econometric models. But so far the forecast accuracy of these models is disappointing, unless the projections are adjusted subjectively by the modelers. In many cases, forecasts by simultaneous-equation (econometric) models "were beaten by those of a simple AR(4) model" (Granger, 1986). Moreover, such AR models themselves do not outforecast naive extrapolation models (Makridakis and Hibon, 1984). Makridakis's conclusion was drawn from an exhaustive comparison of the models on 111 time-series data; the study provides strong evidence that naive extrapolation is all we need in economic forecasting. But then it is natural to ask the next question: What is the difference between a naive extrapolation model and no model at all?

In one study (Dreman, 1984), professors at Harvard and MIT kept track of stock forecasts made by at least five or six analysts. The numbers of companies monitored were 769 in 1977 and 1,246 in 1981. The conclusion was that what forecasters had been touting simply did not pay off. The dismal performance of forecasting models was well summarized by Dreman: "Astrology might be better!"

This situation resembles the often aberrant predictions in geology. People who are desperate for forecasting usually put blind faith into anything associated with sophisticated mathematical formulas. However, a lesson I have learned during years of time-series modeling is this: Forecasting is too important to leave to time-series analysts.

In short, current statistical forecasting simply does not live up to its catchy title. The wholesaling of such techniques to general scientists or to ordinary users will eventually result in mistrust of our profession. For example, I would recommend for the reader's attention an article entitled "Statistical Forecasting Made Simple, Almost" in *The New York Times*, C4, June 24, 1986. One of my colleagues (an influential administrator) was excited about it and was planning to purchase a site license so that we could predict student enrollments and plan better for course offerings. The administrator called me in and asked how many copies of the software we should buy. "Zero; the software is a dud," I replied, showing him a demo-diskette that I had bought one year earlier. The *Times* article was, in my judgment, an outrageous promotion of state-space modeling. With great dismay, I crossed out the title of that article and replaced it with the following: "Statistical Forecasting Made Simple? *Never!*"

III.   REGRESSION AND TIME-SERIES ANALYSIS: SCIENCE OR LUNACY?   (Part II)

An important distinction was proposed by Dempster (1983) to contrast *technical statistics* and *functional statistics*. Dempster made it clear that technical statistics deals exclusively with tools and concepts that have been applied widely across many professions and sciences, while functional statistics requires blending adequate skill and depth in both technical statistics and substantive issues. The two sometimes overlap, but often technical statistics simply do not function.

Dempster, an eminent Harvard statistician, observed that formal analysis of the connection between technical statistics and the real world is "almost nonexistent" and that "very little of this is taught to statisticians." But when scientists need advice on data analysis, whom do they turn to? Further, when research journals are littered with misleading statistics, is it the scientific community or the statistical community that should be held responsible?

EXAMPLE 1   In the spotlight here is a user's guide called *SAS For Linear Models* (Freund and Littell, 1985/1981, pp. 92–100), a very popular "guide" among industrial and academic statistics users. In this case, the authors used the General Linear Model to analyze the test scores made by students in three classes taught by three different teachers. The sample sizes of the three classes were 6, 14, and 11. Randomness of the data was, as usual, not indicated. With this kind of data, stem-and-leaf plots are about the only techniques that are appropriate. Yet the authors spent 8 and 1/2 pages on lengthy discussions of the technical statistics, which include the general linear model, the matrix operations, the estimation of coefficients, the one-way design and balanced factorial structure, the adjusted and unadjusted means, the uniqueness of the estimates, and so on and so forth.

Nowadays statistics users are preoccupied with grinding instead of intellectual reasoning. This kind of practice gives my profession a bad name, and famed statistics educators like Freund and Littell owe us much explanation for the misleading example they have presented to novice statistics users.

In a private conversation, a SAS sales representative defended the example by saying that the authors were trying to show statistics users a valuable algorithm, not a causal link. I responded: "If Freund and Littell are short of *any* meaningful example, they should simply label the data sets as A, B, C, treating them like pure digits." Statistical comparison is a powerful tool for data analysis, but spending 8 and 1/2 pages on the printout of a General Linear Model for students' test scores is both an abuse of statistics and an abuse of the computer as well.[4]

EXAMPLE 2   Ever since its early development, statistics has frequently been called upon to answer causal questions in nonexperimental science. As pointed out by Glymour (1987, p. 3):

> Today, statistical methods are applied to every public issue of any factual kind. The meetings of the American Statistical Association address almost every conceivable matter of public policy, from the *safety of nuclear reactors* to the *reliability of the census*. The attempt to extract causal information from statistical data with only indirect help from experimentation goes on in nearly every academic subject and in many nonacademic endeavors as well. Sociologists, political scientists, economists, demographers, educators, psychologists, biologists, market researchers, policy makers, and occasionally even chemists and physicists use such methods. [emphasis supplied]

Glymour did the statistical community a good service in giving us a clear picture of how popular statistical causal-inferences are in nonexperimental sciences. The popularity, however, does not by itself establish the soundness of the applications of these methods. In the case of the safety of nuclear reactors, the reader is urged to consult Breiman (1985, pp. 203–206) for an insider's account of statisticians' failure on this issue. For the heated debate on the reliability of the census, an illustrating source is Freedman and Navidi (1986).

It is interesting that Glymour cited the meetings of ASA (American Statistical Association) to support his advocacy of drawing causal inferences from correlation data.[5] Anyone who frequents professional gatherings would testify that the presentations in these meetings cannot all be taken as serious scientific investigations. This phenomenon occurs not only in ASA meetings but also, I suspect, across *all* academic disciplines.

In my observation, numerous applications presented in the ASA meetings are like that of Freund and Littell in Example 1: the logical link between the real-life data and the technical statistics presented is simply not there. A notable example is the crude application of statistical models in legal action involving issues of gender and racial discrimination in salaries. In the case of sex discrimination, the following variables are often used:

Y = salaries of the employees in a certain institution

G = 1, if the employee is male

    = 0, otherwise

X = a vector of covariates such as educational background, seniority, etc.

e = error term.

A "direct" regression model of the discrimination is:

$$Y = a + b * G + c * X + e. \tag{1}$$

Some experts in the field also use the following "reverse" regression (see, e.g., Conway and Roberts, 1986):

$$X = a + b * G + c * Y + e. \tag{2}$$

Sex discrimination is then alleged if the coefficient b in expression (1) or (2) is significantly different from zero.

These practices are now common and recently drew criticism from hardline statisticians. In a special session of the ASA Annual Meeting, Holland (1988) discussed the weakness of these models in the context of Rubin's model and concluded that both direct regression and reverse regression provide only sloppy analysis in the case of sex discrimination.[6]

In a similar attack on the use of regression models for employment discrimination, Fienberg (1988, *Statistical Science*) mentioned a case history involving hiring, compensation, and promotion:

When the case went to trial, the centerpiece of the plaintiff's argument was the testimony of a statistical expert who carried out multiple regressions galore.... She concluded that the coefficient for gender was significantly different from zero and proceeded to estimate the damages....

To rebut this evidence the defendant put on the stand a statistical witness who...added a term...to the models of the plaintiffs' expert and noted the extent to which the estimated gender coefficient changed.

The judge, in his written opinion, stated that neither of the experts' regressions had anything to do with the realities of the case, but ruled in favor of the plaintiffs nonetheless.

Appalled by the statistical installations in this case study, Fienberg asked the defense attorney:

Why did the company's lawyers allow their expert to present such mindless regression analyses in response to the equally mindless ones of the plaintiff's expert?

The defense attorney responded:

You don't understand. If the plaintiffs' expert hadn't been busy running multiple regressions she might have taken a closer look at the employee manual which describes what in essence is a two-tiered job system....

When our expert responded by running his own regressions, the lawyers were quite pleased. They believed that the outcome would have been far worse if he had explained to the court what we really do because then the judge could easily have concluded that our system was discriminatory on its face.

As an expert in the field, Fienberg wrote,

Those of us with interests in the legal arena continue to look with horror on the ways in which statistical ideas and methods are misused *over and over again* by expert witnesses (often not statisticians or even those trained in statistics) or misinterpreted by judges and juries. [emphasis supplied]

In his resentful comment on the misuse of econometric models for employment discrimination, Fienberg concluded that a statistician will be better off "to focus on employment process decisions and to learn about the input to those decisions."

In the 1988 ASA Annual Meeting, Holland challenged the audience with the following question: "Is sex a cause in salary discrimination?" Before he gave his answer, Holland quipped: "Sex probably is the greatest of all causes." But in the case of salary equity, Holland maintained that being a man or a woman does not cause the salary inequalities, but that "it is discrimination that is the cause." Bravo, Holland.

But how are we going to assess the magnitude of the discrimination? Holland's answer is downright honest: "I don't know."[7]

EXAMPLE 3   The third example in this section is taken from a book entitled
Applied Time Series Analysis for the Social Sciences (McCleary and Hay,
1980, pp. 244–270). This example reflects very much the way statistical analyses were performed by the expert witnesses in Fienberg's case study.

In an application of time series to population growth, McCleary and Hay
bent on a lavish investment of energy in technical statistics, which include
prewhitening, ACF (autocorrelation function), CCF (cross-correlation function), residual analysis, diagnostic statistics, and so on and so forth. Finally the
authors derived the following multivariate time-series model:

$$p_t = \frac{(.23)(1 - .50B)}{1 - .62B} f_t - \left[\frac{(.23)(3.6)(1 - .50B)}{(1 - .62B)(1 - .44B)} - .952 \right] h_{t-1}$$

$$+ \frac{a_t}{1 - .26B}$$

where p_t is the annual Swedish population; h_t is the harvest index series; f_t is
the fertility rate; a_t is the white noise; and B is the backward operator.

After a 26-page Herculean effort to build the above "formidable model"
(quoted from McCleary and Hay, 1980), the authors devoted only two lines to
the model interpretation: "The model shows clearly that, during the 1750–
1849 century, the harvest had a profound influence on population growth."[8]

Naive statistics users often perceive number-crunching as the ultimate goal
of scientific inquiry. The above examples are only the tip of an iceberg. (The
reader will find more examples in Freedman, 1987.) As a rule of thumb, often
the more computer printouts empirical studies produce, the less intellectual
content they have.

EXAMPLE 4   In the 1960s, London suffered a smoke episode that was
estimated to have claimed several thousand lives. Data from such episodes and
from persistently high smoke areas have been compiled over many years.
These data were available to U.S. health officials, who follow a routine procedure to calibrate the current standard of clean air. The procedure contains a
document that summarizes all available relevant information. The document
was submitted for review and written public comments, before it became final.

In this U. S. study, the British data were useless for an obvious reason: The
health effects observed took place at levels that were much higher than the
U. S. standard. Also, there is no reliable way to convert the British data into
United States measurement, because at that time scientists on the two sides of
the Atlantic used totally different methods to measure the amount of smoke in
the air.

Given this kind of data, no statistical magic can produce useful information.
But according to Breiman (1985, p. 208), the written public comments con-
sisted of over 200 pages. Well over half of this text was consumed by discus-

sions of statistical techniques, such as the correct way to model with multiple time series, transformation of variables, lags, standard errors, confidence intervals, etc.

The technical statistics in these reviews were impressive but totally irrelevant. Breiman refers to these statistics as an "edifice complex," denoting the building of a large, elaborate, and many-layered statistical analysis that covers up the simple and obvious. In this specific example, the edifice building was, according to Breiman (1985), a less admirable product of "a number of very eminent statisticians."

IV.   REGRESSION AND TIME-SERIES ANALYSIS: SCIENCE OR LUNACY?   (Part III)

In 1912, an American statesman, Elihu Root, won the year's Nobel Peace Prize. The Nobel laureate was also widely considered as one of the ablest lawyers the United States has ever produced (the *New York Times*). He once wrote:

> About half the practice of a decent lawyer consists in telling would-be clients that they are damned fools and should stop.

In my observation, many statisticians follow a completely different tactic: They are enthusiastic about promoting the use of statistics and are often reluctant to tell scientists (or would-be scientists) about the severe limitation of statistical inference. Many veteran statisticians are aware of this problem, but they either are busy in producing esoteric articles or simply do not want to rock the boat.

A prominent exception is Professor David Freedman of the University of California at Berkeley. In a series of articles, Freedman (see the references at the end of this chapter) mounts a full-fledged attack on law-like statistical models that rely on regression and/or time-series techniques. The models are among the dominant research methodologies in econometrics and social science research. One of the models under fire was the celebrated study by Blau and Duncan (1967) on the American occupational structure (Freedman, 1983). Blau and Duncan's classic work was cited by the National Academy of Sciences (McCAdams et al., 1982) as an exemplary study in social science.

"Despite their popularity," Freedman (1987, p. 101) declared, "I do not believe that they [the statistical models in question] have in fact created much new understanding of the phenomena they are intended to illuminate. On the whole, they may divert attention from the real issues, by purporting to do what cannot be done—given the limits of our knowledge of the underlying processes." Freedman (p. 102) also asserted: "If I am right, it is better to abandon a faulty research paradigm and go back to the drawing boards."

Freedman (1987, p. 217) insists that multiple regression, as commonly employed in social science, "does not license counterfactual claims." His objections to these models can be summarized as follows. First, the models are devoid of intellectual content. The investigators do not derive the models from substantive knowledge; instead, the models are either data-driven or simply assumed (1983, p. 345).

Second, nobody pays much attention to the stochastic assumptions of the models. In most social-science applications, these assumptions do not hold water. Neither do the resulting models (1985). Some investigators are aware of the problem of the stochastic assumptions in their models and therefore label the computer outputs as merely *descriptive statistics*. "This is a swindle," Freedman declared. "If the assumptions of the regression model do not hold, the computer outputs do not describe anything" (1985, p. 353).

Third, statistics as a science must deal explicitly with uncertainty. But in practice, complicated statistical models often are the dominant *source* of uncertainty in a serious investigation (Freedman and Zeisel, 1988). Instead of solving problems in real life, such models sweep the problems under the carpet.

Fourth, these models do not render meaningful predictions; they only invite misinterpretations (1987). In comparison to natural science models (e.g., Newtonian mechanics and Mendelian genetics), social science models do not capture the causal relationships being studied. In sharp contrast, the natural science models work "not only by log likelihood criteria, but for real" (1987, p. 220).

Freedman's unflattering attack (1987) on social-science statistical models attracted the attention of 11 scholars. Some lent their support, while others expressed their dismay. In response, Freedman wrote:

> We have been modeling social science data since the 1930s, taking advantage of modern data collection and computing resources. The bet may have been a reasonable one, but it has not paid off. Indeed, I have never seen a structural regression equation in economics or sociology or education or the psychology of drug addiction, to name a few fields (not entirely at random). If the discussants have examples, they have been too shy to disclose them.[9]

Freedman's (1987) criticism of the poor performance of statistical models in social (and economic) science is similar to that of Richard Feynman, a famous physicist who placed much of nonexperimental science in the category of "cargo cult science." After World War II was over, according to Feynman, certain Pacific islanders wanted the cargo planes to keep returning. So they maintained runways, stationed a man with wooden headphones and bamboo for antennas, lit some fires, and waited for the planes to land (*The New York Times*, February 17, 1988).

It is the same, Feynman said, with cargo cult scientists. "They follow all the apparent precepts and forms of scientific investigation, but they are missing something essential because the planes don't land."

Social science researchers may feel that physicists and hard-line statisticians are too critical. But as a friend of mine (a psychologist) once told me, "In natural science, people step on each other's shoulders. But in my field, we step on each other's faces."

Despite all these negative comments, social science remains indispensable. In such fields, there are plenty of informative studies (see examples in Freedman, 1987, p. 206; Freedman et al., 1978, pp. 41–43). But these useful studies are based on good measurement, insightful analysis, and simple statistical methods that emphasize the raw data. Sophisticated statistical models, by contrast, tend to overwhelm common sense (Freedman, 1987) and misdirect the course of the research. Good design and measurement are needed in order to develop good theory. "Pretending that data were collected as if by experiment does not make it so." In conclusion, Freedman (1987) wrote:

> Investigators need to think more about the underlying social processes, and look more closely at the data, without the distorting prism of conventional (and largely irrelevant) stochastic models.

Amen.

V. STATISTICAL CORRELATION VERSUS PHYSICAL CAUSATION

Dempster (1983) suggested that although statistical model-building makes use of formal probability calculations, the probabilities usually have no sharply defined interpretation, so the whole model-building process is really a form of exploratory analysis.

Dempster is quite right, and I think it will be healthier if we take away the aura of these models and use them as just additional tools of Tukey's EDA. (The reader will find a similar conclusion in Freedman, 1987, p. 220.) Some researchers may feel insulted and fear a loss of prestige if their models are labeled as EDA; but this is a hard reality they will have to swallow.

A popular conception is that EDA does not involve probability models or complicated formulas. This is entirely mistaken. Consider Tukey's Fast Fourier Transform. The transform is highly sophisticated and widely used in scientific applications. But its main purpose is simply to hunt for hidden patterns, a typical activity of EDA.

As is well known in the history of mathematics, algebra can often yield more than one expects. Therefore, not all statistical modeling that involves intricate algebraic manipulations should be considered useless.

To illustrate, let's first consider the celebrated Wolfer sunspot numbers. The plot of the sunspot data (from 1760–1970) is shown in Fig. 4. This plot exhibits cycles ranging from 8 to 17 years, with an average of approximately 11 years.

An obvious approach to modeling the sunspot data is to fit a sum of sine waves plus a series of random noise. This method follows the tradition of discovering deterministic patterns that were contaminated by noise. The approach is sensible, but it ends up estimating cycles with "fixed frequencies," which were simply not there.

After years of reflection, Yule (1927) abandoned the concept of deterministic models. Instead, he proposed the following stochastic model to fit the sunspot data:

$$X(t) = 1.34X(t - 1) - .65X(t - 2) + e(t),$$

where $X(t)$ is the sunspot number at time t, and $e(t)$ represents white noise. The model is called an AR(2), second-order autoregressive model. This model contains no deterministic components and represents a major philosophical departure from the old paradigm.

A question is: Is Yule's model reasonable? Or more fundamentally: how can Yule's simple model explain the 11-year cycle of the sunspot data?

These questions can be answered by a technique called the spectral density function (Priestley, 1981, pp. 282, 550). For Yule's model, the spectral density function has a peak at

$$f = (1/2\pi) \arccos (1.34/2\sqrt{.65}) = .093$$

which translates into an average period of 1/f, or 10.75 years. Yule's model therefore faithfully indicates an 11-year cycle of Wolfer sunspot numbers.

It should be noted that Yule's model is far from perfect for at least two reasons: (1) The model tells us nothing about physical reasons for the 11-year cycle. In fact, the model is based on statistical correlations, not physical causation. Scientists have tried very hard to understand the mechanism of sunspots

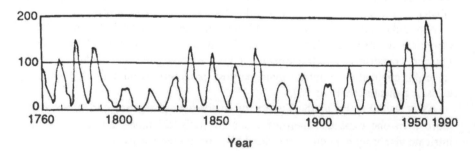

Figure 4   Wolfer sunspot data.

(see, e.g., Dicke, 1985; see also the long lists of references in Bray and Loughhead, 1964). Yule's model is only a black box, concerned with smoothing the curve, rather than gaining a true understanding of the underlying mechanism. (2) The prediction of the sunspot numbers by Yule's model is very poor. The forecast function is essentially a horizontal line with huge prediction errors. Anybody with a ruler can do a better job, especially where the timeplot direction has just turned from upward to downward, or vice versa. (For another discussion of AR models for sunspot data, see Woodward and Gray, 1978.)

Nevertheless, Yule's model remains popular among certain astrophysicists. The model has further become a new research tradition under the name "Box-Jenkins time-series analysis." Like any statistical tool, time-series analysis delivers mixed blessings. It should not be blindly approved, but neither should it be condemned out of hand. A favorable application of time-series models will be shown in the following subsection.

•   •   •   •   •   •   •   •   •

In principle, a good understanding of the underlying mechanism is desirable. Yet in practice the understanding may not help for the prediction or planning. An interesting example in this regard involves the prediction of the geyser eruption in Yellowstone National Park. Many visitors to the park believe that Old Faithful erupts with time-clock regularity. But in reality Old Faithful is not all that faithful: Eruptions may occur within time intervals ranging from 33 minutes to as long as 148 minutes. Presented in Figs. 5(a)-(c) are some frequency distributions of the intervals (Rinehart, 1980).

Upon close inspection, the histograms appear normal only occasionally (in the years 1922, 1923, 1953); and a curious bimodal pattern seems to be the characteristic of the geyser's behavior. Imagine that you are betting on the average interval and that you will win $100 if the difference between the predicted and actual values is less than 10 minutes. In 1923, the frequency of winning would have been about 85%, but in 1963 it dropped to a miserable 25%.

Until 1938 all predictions of Old Faithful were based on the average interval. An important discovery was made by ranger–naturalist H.M. Woodward in 1938: There is a positive correlation between the duration of eruption (X) and the length of interval between eruptions (Y). The striking relation between X and Y can be seen in the time-order plots of Fig. 6 (Denby and Pregibon, 1987). You may have noticed an odd pattern in Day 7. For a marvelous discussion of Day 7, see Denby and Pregibon (1987). Here we concentrate on the seesaw of the timeplots, which reveals that long intervals alternate with short intervals in a fairly regular fashion, a phenomenon closely related to the bimodality in the distribution of Y.

A scatter diagram of Y versus X is shown in Fig. 7 (Denby and Pregibon, 1987).

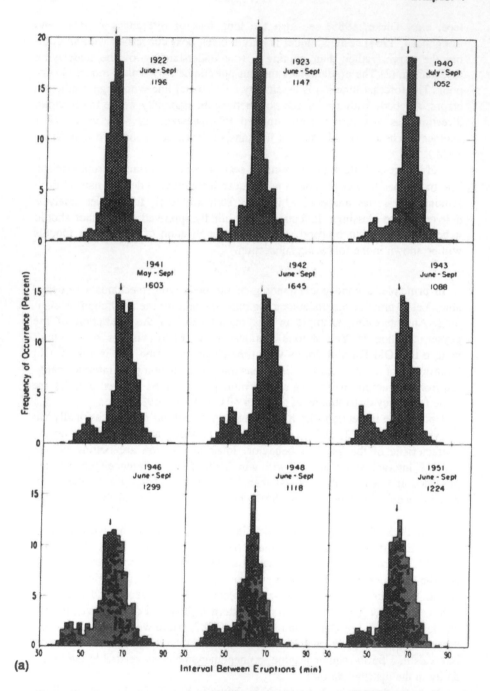

(a)

Figure 5   (a)-(c). Frequency distribution by year for intervals between eruptions of Old Faithful. Copyright © 1980 by Springer-Verlag. Reprinted by permission.

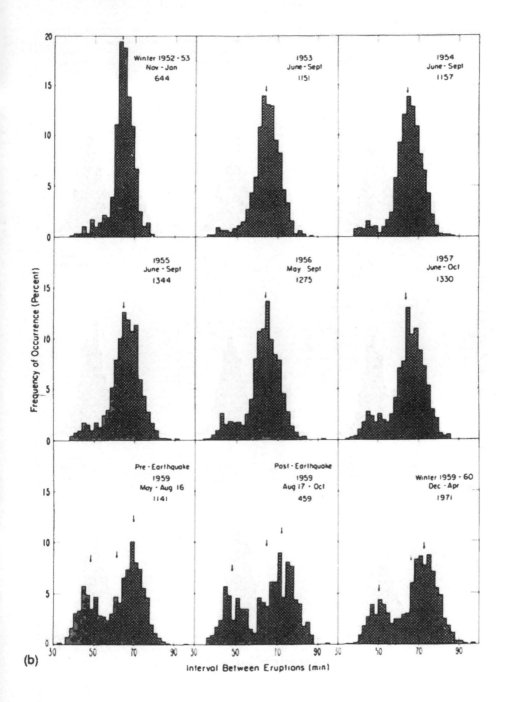

(b)

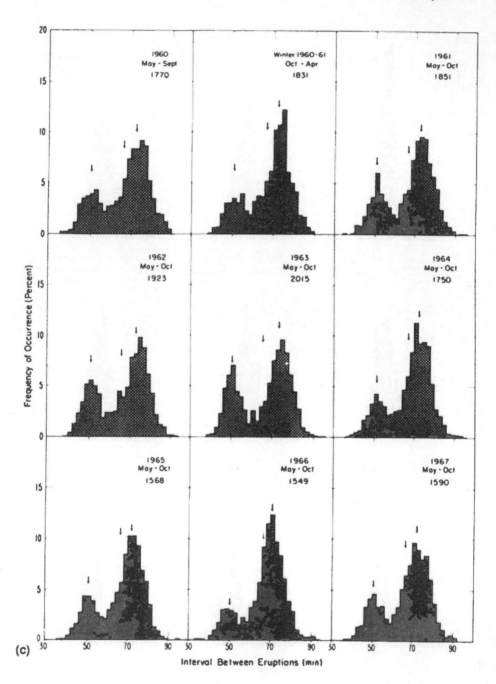

(c)

Figure 5 (Continued)

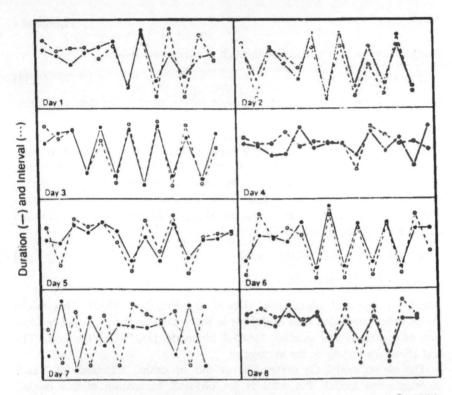

Figure 6 Time-order plot of duration and interval for each day. Copyright © 1987 by the American Statistical Association. Reprinted by permission.

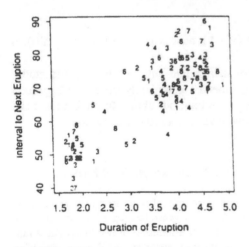

Figure 7 Enhanced scatterplot of interval vs. duration after adjustment and realignment. The variable "day" is used as the plotting character in the plot. Copyright © 1987 by the American Statistical Association. Reprinted by permission.

A regression model passes a line through the above scatterplot:

$$Y = 42.5 + 7.2 * X \tag{1}$$

But what is the relevance of the equation to the geophysical theory of eruptions? A plausible argument (Rinehart, pp. 150- 151) is that a long play empties more of the underground tank than a short play, requiring a longer time for the water rising in the tank to develop sufficient energy for the next eruption. The argument also explains the bimodality of intervals between eruptions.

Note that statistical techniques can be taken beyond the above regression. In equation (1), let e(i) be the residual of the i-th observation. A Durbin-Watson test (d = 3.37) indicates that a serial correlation exists among {e(i)}. The data thus can be refitted by the following model:

$$Y(i) = 38.9 + 8.26 * X(i) + e(i), \tag{2}$$

$$e(i) = (-.5) * e(i - 1) + u(i),$$

where u's represent white noise (Denby and Pregibon, 1987). This model yields better prediction than model (1) in the sense that the predicted residual sum of squares is now smaller: 10665.4 for model (1), 5915.3 for model (2), and 15984 for betting on the average.

The above model (2) makes use of the smoothing technique of Yule's autoregressive model. But what is the physical foundation of this model? Mathematically, model (2) entails the next model:

$$Y(i) = 58.35 + 8.26 * X(i) + 4.13 * X(i - 1) \tag{3}$$

$$- .5 * Y(i - 1) + u(i),$$

which implies that $X(i - 1)$ and $Y(i - 1)$ have a left-over effect on $Y(i)$, a very plausible conclusion.

In this example, scientists cannot manipulate any variables in the study. Neither does model (3) actually describe the complexity of the underground plumbing system of Old Faithful. Nevertheless, our ability to predict has been greatly improved; and next time if you have to bet, you can lose less.

NOTES

1. The Pythagorean regression was inspired by an example in FPP (1978, p. 195–198).
2. The Pythagorean Theorem is also the cornerstone of most statistical procedures: The standard deviation of a batch of data is a generalization of the Pythagorean Theorem; the standard deviation of the sum (or difference) of two independent random variables is the Pythagorean Theorem in a vector space. Even in regression analysis (and ANOVA, the analysis of variance), the Pythagorean Theorem appears in the following statement: "Total variation is the sum of explained variation and unexplained variation."

3. The difference between the coefficients and unity are statistically significant (which may be an indication that the original drawings of the triangles are not really orthogonal).

4. This example does not imply that SAS is a bad software package. Rather, the moral is that even a good package may not provide a safety net for a poor study.

5. These meetings are vital to the professional growth both of the ASA and of individual participants—no doubt about it.

6. This conclusion is obvious if one's common sense is not overwhelmed by the algebra in the models.

7. As an example related to expert witnesses, Neufeld and Colman (1990, *Scientific American*) criticized the use of DNA evidence in forensic testimony. In one case, Lifecodes Corporation (one of the two major commercial forensic DNA laboratories in the U.S.) reported the odds of a random match between a bloodstain and the suspect at 1 in 100 million. On the other hand, Eric Lander of Harvard University and MIT examined the same data and arrived at odds of 1/24.

    In laboratories, DNA analysis can be as reliable as a person's finger prints. Outside laboratories, however, it is hard to fathom the effects of environmental insults (such as heat, humidity, and light) to samples of blood, hair, or semen.

    As a matter of fact, a recent report by the National Academy of Sciences says that the DNA evidence should not be allowed in court in the future unless a more scientific basis is established (*The New York Times*, April, 14, 1992).

8. McCleary and Hay might argue that they simply wanted to demonstrate a statistical algorithm. But perhaps then they would have to change the title of their book from "Applied Time Series" to "Mechanical Time Series" or something of that order.

9. In at least two instances, the performance of statistical models appeared very successful. The first example (Bayesian time-series model) is documented in Chapter 6, Section IV. The second example is an econometric forecasting equation developed by Ray C. Fair, an economist at Yale University. In the last 19 presidential elections, the Fair equation made only three wrongs calls (prediction errors = .9%, 1.7%, .5% in 1960, 1968, 1976, respectively). For the 1992 election, the Fair equation predicts that Mr. Bush will win 57.2 percent of the popular vote in November (*The New York Times*, March 21, 1992). But at the moment of preparing this monograph (July, 1992), all polls indicate that Mr. Bush is slightly lagging behind Bill Clinton (28% vs. 26%, CNN and *Time*, July 12, 1992; 42% vs. 30%, *Washington Post*, July 15, 1992).

REFERENCES

Armstrong, J. S. (1984/1978). Forecasting with Econometric Methods: Folklore versus Fact. *J. of Business*, Vol. 51, 565–593.

Blau, P. and Duncan, O. D. (1967). *The American Occupational Structure*. Wiley, New York.

Box, G. E. P. and Jenkins, G. M. (1976). *Time Series Analysis: Forecasting and Control*. Holden-Day, San Francisco.

Bray, R. J. and Loughhead, R. E. (1964). *Sunspots*. Dover, Mineola, New York.

Breiman, L. (1985). Nail Finders, Edifice, and Oz. *Proceedings of the Berkeley Confer-*

ence in Honor of Jerzy Neyman and Jack Kiefer, Vol. 1, 201–212, Wadsworth, Belmont, California.

Conway, D. A. and Roberts, H. V. (1986). Regression Analysis in Employment Discrimination Cases. In *Statistics and the Law* (M. H. DeGroot, S. E. Feinberg and J. B. Kadane, eds.), 107–168. Wiley, New York.

Dempster, A. P. (1983). Purpose and Limitations of Data Analysis. In *Scientific Inference, Data Analysis, and Robustness*, edited by G. E. P. Box, T. Leonard, and C.F. Wu. 117–133. Academic Press, New York.

Denby, L. and Pregibon, D. (1987). An Example of the Use of Graphics in Regression. *The American Statistician*, Vol. 41, No. 1, 33–38.

Dicke, R. (1985). The Clock in the Sun. *Science*.

Dreman, D. (1984). Astrology Might be Better. *Forbes*, March 26, pp. 242–243.

Feinberg, S.E. (1988). Comment on "Employment Discrimination and Statistical Science," by A.P. Dempster. *Statistical Science*, Vol. 3, No. 2, 190–191.

Freedman, D. A. (1981). Some Pitfalls in Large Econometric Models: A Case Study. *J. of Business*, Vol. 54, No. 3, 479–500.

Freedman, D. A. (1983). A Note on Screening Regression Equations. *American Statistician*, Vol. 37, 152–155.

Freedman, D. A. (1983). Structural-Equation Models: A Case Study. Tech. Report No. 22, U. C. Berkeley.

Freedman, D. A. (1983). Comments on a Paper by Markus. Technical Report No. 25, U. C. Berkeley.

Freedman, D. A. (1985). Statistics and the Scientific Method. *Cohort Analysis in Social Research*, edited by W.M. Mason and S.E. Fienberg, 343–390. Springer-Verlag, New York.

Freedman, D.A. (1987). As Others See Us: A Case Study in Path Analysis. *J. of Educational Statistics*, Vol. 12, No. 2, 101–128, 206–223; with commentaries by 11 scholars, 129–205.

Freedman, D. A. and Daggett, R. S. (1985). Econometrics and the Law: A Case Study in the Proof of Antitrust Damages. *Proceedings of the Berkeley Conference in Honor of Jerzy Neyman and Jack Kiefer*, Vol. 1, 123–172. Wadsworth, Belmont, California.

Freedman, D. A. and Navidi, W. C. (1986). Regression Models for Adjusting the 1980 Census. *Statistical Science*, Vol. 1, No. 1, 3–11, with commentaries 12–39.

Freedman, D. A., Navidi, W., and Peters, S. C. (1987). On the Impact of Variable Selection in Fitting Regression Equations. Technical Report No. 87, U. C. Berkeley.

Freedman, D. A., Rothenberg, T., Sutch, R. (1982). A Review of a Residential Energy End Use Model. Technical Report No. 14, U. C. Berkeley.

Freund, R.J. and Littell, R.C. (1985/1981). *SAS for Linear Models: A Guide to the ANOVA and GLM Procedures*. SAS Institute.

Glymour, C., Scheines, R., Spirtes, P., and Kelly, K. (1987). *Discovering Causal Structure: Artificial Intelligence, Philosophy of Science, and Statistical Modeling*. Academic Press, New York.

Granger, C. (1986). Comment on "Forecasting Accuracy of Alternative Techniques: A Comparison of U.S. Macroeconomic Forecasts," by S.K. McNees. *J. of Business and Economic Statistics*, Vol. 4, No. 1, 16–17.

Holland, P. W. (1987). Causal Inference and Path Analysis. Presented at the 1987 ASA Annual Meeting, San Francisco.

Holland, P. W. (1988). Causal Inference in Salary Discrimination Studies. Presented at the 1988 ASA Annual Meeting, New Orleans.

Klein, L. R. and Goldberger, A. S. (1955). *An Econometric Model of the United States, 1929-1952.* North- Holland, Amsterdam.

McCAdams et al. (1982; ed.) *Behavioral and Social Science Research: A National Resource.* National Academy Press, Washington, D.C.

McCleary, R. and Hay, R. A. (1980). *Applied Time Series Analysis for the Social Sciences.* Sage, Beverly Hills, California.

Makridakis, S. and Hibon, M. (1984/1979). Accuracy of Forecasting: An Empirical Investigation (with Discussion). *J. of Royal Statistical Society, Series A*, Vol. 142, Part 2, 97-145.

Priestley, M.B. (1981). *Spectral Analysis and Time Series.* Academic Press, New York.

Rinehart, J. S. (1980). *Geysers and Geothermal Energy.* Springer-Verlag, New York.

Woodward, W. A. and Gray, H. L. (1978). New ARMA Models for Wolfer's sunspot Data. *Commun. Statist.*, *B7*, 97-115.

Yule, G. U. (1927/1971). On a Method of Investigating Periodicities in Disturbed Series with Special Reference to Wolfer's Sunspot Numbers. In *Statistical Papers of George Undy Yule.* Hafner Press, New York.

Intermission

In the history of science, the single most important breakthrough is probably the publication of Newton's *Principia* (1687). This body of work marked the birth of calculus and its applications to physics, astronomy, and to numerous branches of natural sciences.

In the quest to develop "social physics" and "natural laws of society," researchers in the soft sciences have found that calculus is not readily applicable. In addition, nowhere can a grand unification like Newton's theory be anticipated. Researchers thus turn to a suspect area of mathematics that is considered as a "social calculus" (or "psychology calculus") but is more widely known as statistics.

It is fair to say that researchers in non-experimental science are the major consumers (and abusers) of statistical methods. In these branches of sciences, often there is little theory in their inquiry. Therefore data compilation and statistical manipulation are the dominant components of their research "methodology." In such fields, statistics are mass produced and frequently misinterpreted. Worse, researchers in such fields often disapprove of subjective knowledge as non-scientific." Such researchers are very adamant about the "rigor" and the "objectivity" of their statistical results. To them, this is what they have learned from statistics textbooks and from consultation with famous statisticians. When pressed, these researchers often retreat to saying that "social-science studies are very complicated," implying that problems in natural science are cut-and-dried and easier to manage. This assertion is definitely not true, and we will give living proofs from natural science to discredit this belief.

It should be made explicit that we have no intention to look down at soft sciences such as social, behavioral, and economic research. To clarify this position, we would like to quote a study reported by the National Research Council, an arm of the National Academy of Science. The report compared the length of time doctoral candidates took to complete degrees that were awarded in 1987. Here are the median years spent earning the diploma by doctorates in some selected fields:

Field	Median years
Engineering	5.8
Physical sciences	6.0
Social sciences	7.2
Education	7.9
Humanities	8.4

It is evident that social scientists in general spend more time pursuing their advanced degrees. In principle, there is no reason their research findings should deserve less respect than those in natural science. Nevertheless, one often finds that the work of certain *quantitative* and "objective" social-behavioral scientists is devoid of intellectual content. In sharp contrast, many *qualitative* social scientists are walking encyclopedias in their own right, and it is always an intellectual delight to consult with this kind of scholar.

In the second half of this book, we will try to restore the value of certain traditional (and speculative) methods in social-behavioral research. We will also provide a behind-the-scenes look at the statistical activities in many branches of modern sciences, where the belief of "objective statistical methods" has misguided investigators in otherwise illuminating investigations. Further, we will put forth a synthesis of subjective and objective knowledge. Several case studies will be brought about to support the necessity and the contents of this synthesis.

The second half of this book is thus organized as follows. Chapter 5 consists of two case studies. Section I provides a case study about the sorry state of statistical evaluation in educational research. A conclusion is that anecdotal evidence could be more convincing (and more revealing) than so-called "statistical evidence."

Section II paints a mixed picture of statistical modeling in a social-behavioral study. On one hand, we pause in admiration of a qualitative analysis in a ground-breaking paper (Bem, 1974); on the other hand, we criticize a host of simple- minded statistical models related to the Bem study. We finally condemn a religious practice of statistical "reliability and validity" in psychometric measurements.

In addition, Chapter 5 documents the mistakes, the growth, and the frustration of a mathematical statistician (this author) who wandered into the jungle of statistical analysis in soft sciences. Statisticians who have to deal with soft scientists may find the chapter relevant.

Chapter 6 is devoted to many aspects of objectivity, subjectivity, and probability. To this end, the chapter first criticizes the statistical justification of "scientific philosophy." Ludicrous justifications of "objective knowledge" are similar and abundant in the soft sciences. Such misconceptions of "objective knowledge" are countered, in this chapter, by the Bayesian assessment of subjective probability and subjective knowledge.

Chapter 7 presents scholarly debates over ad-hoc methods versus "systematic" applications of statistical procedures in non-experimental sciences. In so doing, the chapter tries to maintain a balance between quantitative and qualitative analysis, and between order and chaos. Examples from the field of dynamical systems are brought in to support this position.

Chapter 8 is concerned with the curious fact that while statistics appear everywhere, the discipline of statistics lacks visibility and influence on our society (Marquardt, 1987, *JASA*). The discussion of this curious fact is a further attack on the misconceptions of "quantitative analysis" and further supports the synthesis of objective and subjective knowledge.

To enhance the status of the statistical discipline, many statisticians have been tapping the concept of statistical quality control championed by Edwards Deming, the most famous living statistician. Therefore, the second half of Chapter 8 is devoted to discussion of the misconceptions of statistical quality control and to the conflicting teachings of Q. C. gurus.

Finally, we try to put forth a new perspective on statistical inference. The scientific landscape has long been littered with misused statistics. With effort and luck, maybe we can reduce the misuse, 1% per generation.

Chapter 5

A Critical Eye and an Appreciative Mind Toward Subjective Knowledge

I. THE SORRY STATE OF STATISTICAL EVALUATION: A CASE STUDY IN EDUCATIONAL RESEARCH

Since the 1970s there has been a steady decline in the number of doctorates in mathematics, physics, and engineering awarded to Americans by U. S. colleges and universities. In mathematics alone, the number is less than heartwarming. (See Fig. 1, Notices of American Mathematical Society, November, 1986.) The shrinking number of American mathematicians worries the mathematical community. A mathematician on the East Coast believes that if we double the salaries of all mathematicians, then more students will come in to the field of mathematics. On the contrary, famous mathematician Paul Halmos is widely known to have said that we have too many mathematicians, and should therefore cut their salaries by half.

Less controversial is the need for the improvement of mathematics and science education among the young. Dozens of studies have shown the poor performance of U. S. students in science and mathematics (AMSTAT News, March, 1989). These studies include the widely known 1985 report, *A Nation at Risk*, and *Science for All Americans*, a report by the American Association for the Advancement of Science (1989). The latter report criticized the present curricula in science and mathematics for being overstuffed and undernourished. The report generated a milestone, project 2061—a long-term, three-

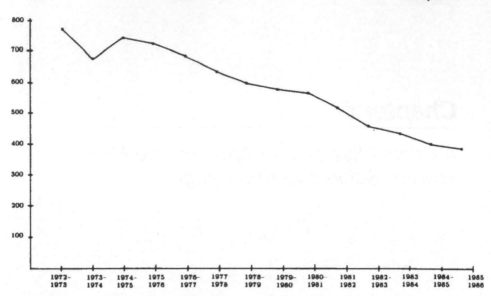

Figure 1   Total number of doctorates in mathematics who are U.S. citizens. Copyright © 1986 by the American Mathematical Society. Reprinted by permission.

phase AAAS initiative. One of the recommendations of the latest report is that students should develop their thinking skills, memorize fewer details, and focus on the connections between subject-matter areas instead of the boundaries.

Of special concern in this chapter is the teaching of calculus courses. To informed academicians, calculus is, by any measure, one of the greatest achievements of western civilization. But how many of our students know this? In this country a typical way of teaching calculus is for the instructor to copy the formulas from the book to the blackboard, just making all the characters bigger. In brief, calculus is presumed to be cold, dreadful, mechanical, and depersonalizing.

At the American Mathematical Society (AMS) centennial celebration in 1988, some mathematicians called the current approach to calculus instruction "a national scandal." These mathematicians contended that if America wants to compete effectively in the world, it is inadequate for just a relatively few elite students to understand calculus. Millions of other students must also comprehend it, so they can fill vital jobs in science, engineering, and technology. The scholars observed that calculus teaching methods have failed to keep up with changing times—which is ironic because calculus is mathematician's primary tool for describing change (The *Philadelphia Inquirer*, November 6, 1988). In the hope of making calculus more appealing and more manageable for students, the National Science Foundation awarded 25 grants in October

1988 as the start of a planned national campaign to revamp calculus instruction.

In January 1989 a mathematical educator at an east coast institution wrote up a proposal for using computer (and graphic calculators) for teaching and learning Calculus I and II (both are required courses for 40% of students on that campus). In the proposal statistical techniques are offered as methods to test the validity of the project, and I was invited to be the statistical consultant "to insure that the design of the experiment will be both valid and reliable."

The project included three instructors, and the design of experiment was described as follows:

> Each of the three faculty members will teach two sections of the calculus each semester, MAT 227 Calculus I in the Fall semester, MAT 228 Calculus II in the Spring semester. For each instructor one section (experimental group) will use a computer approach, the other section (control group) the more traditional paper and pencil approach.

Also, "students in the six sections involved in the project will be given pre- and post tests to measure their growth, with the final examination each semester serving as the common post test."

This design looks like the two-sample comparisons in biomedical research. However, there are pronounced differences: In biomedical research, randomization and double-blind control are used to iron out the collective effects of confounding factors; while in this study such devices cannot be employed.

I told the educator about the potential problems of non-randomness, Hawthorne effect, ceiling effect, the effect due to different hours of study, etc.; but I believed that a careful analysis would give certain credibility to the study.

The statistics that will be coming out of the study are not like clinical trials (where randomization and large samples are used); *rather, they are like a medical doctor's diagnosis* (where a patient walks into his office and the doctor has to make a judgment). In the latter case the doctor measures the patient's blood pressure, height, weight, cholesterol level, blood sample, stool sample, etc. Based on these numbers, the doctor finally develops a prescription (or certain therapy) and hopes that the illness will be ameliorated.

It should be noted that the doctor cannot randomize the patient's blood in order to take a "representative" sample. Nevertheless, he can be more trusted to get a reliable sample and useful information than an apprentice who is thrown into the task. For example, when the doctor is going to take a blood sample, he might instruct the patient not to eat food that may confound the study.

In this calculus project, the statistical consultant cannot simply run a two-sample comparison and then conclude that the computer group is doing better than the paper-and-pencil group. Instead, he has to perform like the medical doctor who uses the best of his knowledge for the benefit of the patient. I considered this fact a big *challenge* to a mathematical statistician: A medical doctor has many years of training and real life experience on a specfic disease (e.g., an eye doctor won't attempt to treat a stomach ulcer); a statistician cannot be an overnight expert on educational research. However, I truly believed that computers would make a difference in the teaching and learning of calculus and probability theory, and I thought if somebody *had to* look at the statistics of this calculus project, I would be a good candidate.

In a hasty manner,[1] I told the educator some of my concern about the reliability of the statistical evaluation. Nevertheless, I gave a nod to the two-sample comparison[2] and agreed to serve as the statistical consultant to the project.

In May 1989 the proposal was fully funded ($25,000 from a state educational fund). Two weeks later I told the director of the project that there were some problems with the statistical design, and that we needed to improve it before the experiment started. In brief, the previous design (the two-sample comparison) had three problems: (1) not scientific, (2) not ethical, and (3) not effective. Other than that, everything was fine.

The director was shocked by my first written report. He said this design is exactly what he had done in his doctoral dissertation at a prestigious university. Furthermore, the design was recommended (and demanded) by a famous statistician who served on his dissertation committee. The director emphasized that the statistician is nationally known and that the two-sample design is used in almost all educational studies.[3]

It appears that the popularity of the method is by itself taken as evidence of scientific merit. Because of this popularity, I sensed that I would have to fight against a wall.

On the one hand, I don't really have a problem with the two- sample design used by doctoral candidates in their dissertations. Most of those dissertations are not funded (or poorly funded) for small-scale artificial studies. On the other hand, this calculus project is different. It has full support from a state government, and it is supposed to make a real and permanent change on that campus. Further, research papers will be generated from this study and submitted for publication. If the results are published, the findings may be cited by proponents as "scientific evidence" for other projects, and by opponents as examples of ludicrous statistics.

After getting feedback from various sources, I wrote a second report to the experimenters. Below are the potential pitfalls of the two-sample design, as well as some exchanges between the experimenters and the statistical consultant.

A. Not Scientific

Before we get into detailed discussion, it should be noted that mathematicians on that campus were sharply divided on the issues. At least one third of the faculty worried that this computer approach would water down the course, and that students would spend more time fooling around in the computer lab than doing real mathematics. Also, they worried that the statistics coming out from the study would be used to discredit the traditional method. Only one third of the faculty were really convinced that this was the right direction to go, and the rest were taking a wait-and-see attitude toward the debate. In short, no honest mathematician can settle for "push-button" calculus, and it will be a formidable task for the proponents of changes to maintain the old standard and to convince the skeptics.

To begin with, skeptics could challenge the comparison on the ground that experimenters are eager or forced ($25,000 at stake) to produce positive conclusions from the experiment. In other words, a major reason why people may not trust the experiment is the instructor's enthusiasm for the new.

For example, students' scores can be easily skewed by (1) higher expectations (or simply a threat of a hard exam), (2) more homework and more hours of study, (3) homework that is more related to test materials, (4) the eagerness of success on the part of the instructors, etc.

An honest experimenter may be able to reduce this bias to a certain degree. But he or she cannot change the fact that people tend to alter their performance when they are under study. This phenomenon is known as the Hawthorne effect (see, e.g., Deming, 1982, p. 109), and it is proportional to the enthusiasm of the experimenters in social-behavioral studies.

Another confounding factor in this calculus project is that many students in the control group may purchase graphic calculators (or computer software) for themselves.

To make the description more vivid, here is a comic version of what may happen in the classrooms: In the first hour (experimental group), the instructor demonstrates to his students the wonderful things that can happen when one is using computer (or graphic calculators) to solve calculus problems; then in the second hour (control group), he tries not to show these things to his students, only to find out that the students are staring at him and that they all have graphic calculators in their hands.

It should be noted that not all confounding factors work in favor of the experimenters. For example, poor students may drop out of the pencil-and-paper group. The dropouts may enhance a "ceiling effect" that usually goes against the validity of the experiment. In other words, with better students and capable instructors in the study, the test scores of both the control and the experimental groups might be so close to the ceiling that any difference between the two groups is of little or no importance, either practically or statistically. This means that even if you have strong evidence that students benefit greatly from the use of computers, it is unlikely you can show it by the statistical comparison of final-exam scores. (This kind of problem is serious in many social-behavioral studies.)

The mathematical assumptions behind the textbook two-sample comparison are very simple:

1. $X_1, \ldots, X_n$ are independent and identically distributed with a finite variance.
2. $Y_1, \ldots, Y_m$ are independent and identically distributed with a finite variance.
3. X_i and Y_j are independent, for all i and j.

These assumptions are relatively easy to meet in biology, engineering, or chemical sciences. In these disciplines population units are almost fixed and rarely interact with each other—conditions hard to maintain in educational settings. For example, how can we guarantee the independence assumptions among X_i's, Y_j's, and between $\{X_i\}$ and $\{Y_j\}$?

Worst of all is that the mathematical model of the two-sample comparison does *not* deal with the randomness and the sources of the difference between X and Y. The difference may be due to a large number of factors (Hawthorne effect, more hours of study, SAT scores, college GPA, majors, you name it).

If randomization is employed at the beginning of the experiment, then we should be able to iron out the effects of SAT, college GPA, etc. But in this study, randomization cannot eliminate the Hawthorne effect, or the effects due to more hours of study, more interaction among students, and between students and teachers, etc. Also, we *want* students to interact. Don't we?

In this calculus project, randomization was not possible in that institution. This fact makes a good design to control variables (that are not eliminated by randomization) very difficult. For example, if the pretest scores are quite different between treatment and control groups, then the adjustment of the posttest scores can be very messy. To sum up, it is likely that one will end up with a "control group" that cannot be controlled at all.

B. Not Ethical

If I were among the students (or parents of these students) in the control groups, I might charge the experimenters with unethical conduct: We paid the same tuition, so why don't we deserve the same education? The grounds of this charge are as follows: (1) The experimenters in this project are positive and enthusiastic about the new approach. You know it is better to use computers, why can't you do it in our class? (2) As discussed above, the comparisons of test scores of experimental and control groups do not really have a sound scientific ground. So why do you have to waste time on the comparison?

Also, if the exam is written so that it does not test "pencil and paper" ability, then the students in the standard method can charge that they did not have a fair test. Further, students in the standard group might charge that the faculty wish to see the experimental group do better, and that therefore the faculty have written a test biased for poor performance by the control group.

With these warnings in mind, the instructors have to prepare themselves to meet with distrustful eyes or even uprisings in the classrooms.

After my first report, one experimenter wrote back: "I am not convinced that computer software or calculators will yield superior results, and if not, so be it." This statement sums up the instructors' responses to my charge of potentially unethical conduct. With their *neutral* attitudes, it also appears that they can escape the Hawthorne effect and produce results that have sound scientific ground.

However, my guess is that if the funding agency had known their attitudes in advance, they would not have had a chance to obtain this grant ($25,000, equivalent to a 3-hour released time for a college professor for 20 semesters). They have received a $25,000 grant. They have to make it work. Don't they?

As a matter of fact, in the proposal (p. 10) the principal investigator stated, "It is *clear* that the use of computers in the classroom is not a 'passing fancy,' but is the reality of the future." This does not sound like a neutral attitude at all.

The indication that the experimenters are not positive toward the experiment is a *bad* sign for any scientific investigation. As far as I know, scientific progress is strongly correlated with the obsession of the investigators in a study. This is especially true for modern science. Can you imagine that a group of scientists who are not convinced of the superiority of a new method will make significant progress in the frontier of human knowledge? This question leads to the next pitfall—that the design is not effective.

C. Not Effective

It should be noted that there is nothing wrong with being enthusiastic about one's experiment—that is the driving force of all scientific progress. What worries people is that one's enthusiasm may not yield a fair comparison.

In contrast to the experimenters, I am openly excited about the use of computers in the teaching of calculus. In my probability course, computers were used both in classroom demonstrations and in homework assignments. It was a new experiment using a new book, a new software, and a new method. Yet students' written evaluations are mostly positive, such as "very interesting," "helps a great deal," "very useful," "fascinating," etc.

In summary, I strongly believe that computers can help to make a real change in the teaching of calculus. But the experimenters have to *find an effective way* to capitalize on the advantage of using computers in the classroom. If the experimenters sacrifice this for the sake of statistical evaluation, then they have gotten their priorities mixed up.

According to the original proposal, the instructors will teach Calculus I in the fall, and then teach Calculus II in the spring. In my judgment this is only a waste of time, because in the classroom an instructor usually teaches better the second time through. For this reason I made the following suggestion: Teach with the new methods to *both* Calculus I *and* II in the fall; then repeat and improve the whole experiment in the spring. If you do only Calculus I in the fall (and then Calculus II in the spring), you will miss out on many opportunities to explore the advantages and disadvantages of using computers in teaching specific topics.

All great scientific results have been repeated over and over; many fundamental results in physics and chemistry are repeated even in high schools by countless would-be scientists. Only certain social scientists do their experiments once, calculate statistical P-values, and then conclude that their results are "scientific."[4]

In summary, the use of the two-sample comparison as described in the proposal is hard to defend morally or intellectually, and the investigators will be better off to abandon it altogether.[5]

D. A Method for Evaluation

One experimenter wrote, "if we eliminate all comparisons, then we can only say we have a case-study, which will leave nothing to base an evaluation upon." This is not true. The internal evaluation described in the proposal has little value, and is indeed harmful to this project and to the purpose of education. For this reason I suggested a method of external evaluation: (i) Recruit instructors (especially young, non-tenured professors) to teach calculus in the spring, using the new methods and guidelines provided by the experimenters.[6] (ii) Collect the test results and written evaluations from these instructors and students by the end of that semester.

If the majority of these instructors endorse the new approach, then the experimenters can cheer and celebrate. Otherwise, few if any will believe their results, no matter how good the test scores of the experimental groups look.

It should be noted that the investigators are lucky to have many colleagues who are neutral or negative toward the calculus project. The experimenters will have achieved a big victory if they win over the skeptics and neutralize the stubborn critics.

More importantly, the extrinsic evaluation has the potential to carry on the mission of this one-year project and make computers a permanent part of calculus instruction on that campus.

Before we go on to other details, let's ask the following question: What is the purpose of this experiment—to spend $25,000 on hardware, software, traveling, and then unplug the whole project one year later?

Of course not. One of the missions of this project is to bring a new life to the teaching of calculus. I believe this can be done, and in fact it has already happened at many other institutions (see, e.g., Heid, 1988, or Sec. I.F. of this chapter). Our experimenters cannot merely say, "I am not convinced that the computer software or calculators will yield superior results, and if not, so be it."

E. Some Technical Details Regarding Statistical Evaluation

A common final exam appears to be a useful benchmark for assessment, but the question is: What should be the content of the exam? Here is a dilemma: If the tests emphasize traditional skills, then the tests are biased against the computer groups; on the other hand, if the tests emphasize computer skills, then they are biased against the traditional groups.

Note that the traditional approach cannot be completely wrong. After all, it is the traditional approach that produced all the scientists, engineers, and mathematicians we have today. Thus it is the experimenters' job to identify the strengths of the traditional method of teaching, and then devise measures for assessment.

A major weakness of the traditional approach to teaching is that students are so overwhelmed by formulas that they seldom try to *reason from basics*. A computer itself cannot solve this problem, and in many cases it may only weaken students' ability to reason from basics. It is therefore another formidable job for the experimenters to correct these shortcomings and to devise measurements for evaluation.

In his original proposal the Director wrote, "Algorithms not used on a consistent basis are easily forgotten, while generalized problem-solving techniques are retained for a much longer time." This conclusion is right on the mark, but needs further elaboration and assessment. Follow-up studies are possible. For example, one could give students a test in the following semester to evaluate the retention of basic concepts in calculus.

To sum up, just like a medical doctor collecting a blood sample from a patient, the investigators can collect sample statistics from the experimental classes for evaluation—final exam, follow-up quizzes, measurements on strength (and weaknesses) of computer and traditional methods, and questionnaires to measure students' interest, expectations, confidence, and appreciation of calculus, etc. If instructors who use traditional methods are willing (or paid) to collect similar statistics, then one can use them for comparison and thus broaden the basis of assessment. All these steps can be carried out without the naive two-sample comparison stated in textbooks.[7]

F. Anecdotal Evidence: A More Convincing Study

A point that has to be made explicitly is that in social-behavioral research, anecdotal evidence can be more useful than so-called "statistical evidence." In other words, statistical evidence is *not* more scientific than anecdotal evidence—it only looks more scientific.

In fact, Heid (1988) conducted a study that is anecdotal in nature, but the results are quite convincing. In this study the subjects were college students enrolled in two sections of a first-semester applied calculus course at a large university. The investigator taught both sections, and used them as the experimental group.

The materials used in these two classes were developed and revised by Heid (instructor A) on the basis of a semester-long pilot study. This is a good start, because the investigator had learned many lessons from the previous work. The approach Heid developed in the teaching of this applied calculus course was quite novel, and may shock most traditional mathematicians. During the first 12 weeks of the course, the students and the instructor used the microcomputer to perform most of the algorithms that were needed in problem-solving; only during the last 3 weeks of the course did the instructor cover traditional procedures (with little attention to applications or to problem-solving).

On the other hand, the control group was a large lecture class taught by another mathematics professor (instructor B). The emphasis in assignments and tests in the control group was the execution of traditional skills, with a small number of concept-oriented questions included on the tests for purposes of comparison with the results from the experimental classes.

The experimental and the comparison classes sat for the same final exam, which was designed by instructor B. Here are some statistics compiled from the final exam:

Final Exam Scores						
Class	n	Mean	SD	25th percentile	Median	75th percentile
Experimental section 1	18	105	43.6	87	105	147
Experimental section 2	17	115	40.9	107	127	145
Comparison section	100	117	36.7	94	116	145

Percent Correct on Individual Final Examination Items				
Item	Content	Experimental section 1	Experimental section 2	Control group
1	Compute derivatives	49	53	53
2	Apply tangents	44	59	44
3	Properties of curves	39	35	37
4	Optimization	17	18	16
5	Exponential growth	50	71	70
6	Compute integrals	53	54	64
7	Area, volume	22	21	28
*8	Optimize f(x,y)	11	12	26
9	Lagrange multipliers	33	24	35

A major difference between the experimental sections and the comparison group is in item 8 (11% and 12% versus 26%). This item required the use of the second derivative test, whose derivation was far beyond the scope of the course. And the students in the experimental classes had been told that memorization of the particulars of that test was not important. Other than item 8, the overall difference between experimental and comparison groups is rather small. The investigator did not report statistical tests, analysis of variance, or type I and type II errors. And fortunately the editor of The *Journal for Research in Mathematics Education* did not bother with the tests of significance that are largely irrelevant, misleading, and over-elaborated in many social-behavioral studies.

The above summary statistics indicate that the students in the experimental and comparison groups performed rather similarly in a traditional test for a calculus course. So what is so good about Heid's new approach?

First, her students spent only three weeks on the test material, while the other students spent 15 weeks. Second, Heid provided a rich array of anecdotal evidence to shed light on the issue under investigation. The evidence was gathered from other sources, such as interviews, a questionnaire that addressed perceptions and attitudes about learning mathematics and calculus, observations in the computer lab and the classroom, student–student interactions, etc.

The most interesting part is the interviews. The procedure was as follows:

> According to their responses to a questionnaire at the beginning of the course, the students were classified into the four categories reflecting their responses (high or low on each) to questions aimed at assessing (a) their need for algorithms and (b) their need for creative work. A stratified random sample of interviewees (stratified on these categories) was selected from among volunteers. All but a few members of the experimental classes and a large number of the comparison class members volunteered for the interviews (monetary compensation was offered).

The interviewees from the experimental classes were similar to those from the comparison class in their average scores on the ours final (experimental average: 137.3; comparison average: 139.8). Some of the interview results will now be described.

1. According to Heid, the interviewees in the experimental classes gave a broader array of appropriate associations when they were asked "What is a derivative?" Also, the wording was often clearly their own:

> A derivative is a way of looking at something.... Knowing the derivative is a rate is something that tells me something intrinsic about the graph.... Now I can look at a graph and say, "OK. The line is going up but how fast is it going up? And is it going down at any point?" And look at just the whole change in the graph, and I can see a lot more.
>
> A derivative is a number of things. Mathematically it's a slope—but I think of it as a rate of change—a description of change. In math, the symbolic language we've been using, we've been working in two variables. The derivative is a description of the

change in one versus the other—at a certain point—at a certain time—at a certain place.

Responses from the students in the comparison class, although not inaccurate, were lacking in the detail provided by the students in the experimental classes:

The change in slope. Change of rate. First derivative. You know—things like that.
It has an explanation of that in the book. I couldn't tell you now exactly what it is. . . . A derivative is useful to find a max or min. I don't know, really.
They explain it as being what the curve is doing—and I don't see the curves. I can think of it as how fast the curve is going. . . . If it's postive, I keep thinking it's definitely going somewhere—if it's zero, it's not—if it's negative, it's going the wrong way.

2. There were occasions during the interviews when the students realized that they had made misstatements about concepts. On many of these occasions, on their own initiative, the students in the experimental classes reconstructed facts about rate of change, concavity, and Riemann sums by returning to basic principles governing the concepts of derivative and area. They frequently used the basic concept of derivative to reason about other concepts: shape of a curve, second derivatives, related changes, concavity, and applications of calculus.

In contrast, when the students in the comparison class verbalized that they had made erroneous statements about concepts, there was no evidence of attempts to reason from basic principles. They often alluded to having been taught the relevant material but being unable to recall what had been said in class.

3. The interviewees from the experimental classes constructed representations of the Riemann sum concept that differed in form from those presented in class or in the text. One student, for example, who conjured up an appropriate trapezoidal rule (not covered in class or in the text) spoke further about the difference in error for upwardly concave and downwardly concave curves. No instances of such constructions were found in the transcripts of the interviewees from the comparison class.

4. The interviewees from the experimental classes verbalized connections between the derivative concept and its mathematical definition, and one interviewee used the proof of the fundamental theorem of calculus to correct a formula he had forgotten. Such instances of using connections between concepts were not found in the transcripts for the comparison group.[8]

The above results of interviews are intriguing. I know of no statistical method that could similarly penetrate into the phenomenon. Further, as a teacher I would be extremely delighted if some of my students (after only one semester of study in calculus) could use the proof of the fundamental theorem of calculus to correct a formula he had forgotten. I would also be elated if some students could conjure up an appropriate trapezoidal rule that was not covered in class or in the text. Isn't this what education is all about?

For stubborn critics, there are plenty of reasons not to buy into Heid's story. For example, the sample size of the experimental group is relatively small (n = 37); Heid might have had better students in the experimental sections; and her personal enthusiasm might have substantially swayed her students toward the learning of calculus.

In order to judge Heid's results on a larger scale, I believe that her approach has to go through a process of natural selection, or the survival of the fittest. In other words, Heid's results will be more convincing if other instructors follow a similar approach and report success (at least to certain degree) in their classroom experiments. On the other hand, her work would have to be deemed a failure if nobody were interested in this method and if Heid herself abandoned the approach a few years after her original publication.

G. Large-Scale Evaluation: A Few Words from an Expert

Some researchers in the field of educational evaluation dislike the process of natural selection in that it is slow and wasteful (Gilbert and Mosteller, 1972). Instead, they promote the use of *true experiments* by arguing their programmatic advantages. Mosteller (1981), for example, has pointed out the value of experimentation by demonstrating the frequency with which innovations turn out to be not as effective as anticipated.

In order to test the treatment effect, a true experiment, according to orthodox statisticians, has to be a randomized and double-blind controlled experiment (Freedman, Pisani, and Purves, 1978, p. 6). But in educational evaluation nobody can really achieve this key ingredient of a true experiment.

Regarding large-scale educational evaluation, let us consider a few words from an expert (Boruch, 1982). Boruch was the main author of a report submitted in 1980 to Congress and the U. S. Department of Education. The report attempted to address issues surrounding the products of the Government's $30–40 million annual investment in educational evaluation.

On the issue of experiments, the report made two recommendations. First, it recommended that pilot studies be undertaken before new programs or new variations are adopted. This recommendation sounds like an old cliché, but "pilot testing is not yet a common practice" (Boruch, 1982).

Second, the report recommended that "higher quality evaluation designs, especially randomized experiments, be authorized explicitly in law" for education testing. By authorization, Boruch (1982) explained, "we mean explicit and official permission. We do *not* mean that such designs should be required uniformly, for such a demand is absurd. Nor do we mean to imply that random-ized tests are the only option in every setting."

Boruch also discussed certain problems in mounting randomized tests. For instance, some 30 randomized tests for Head Start programs were initiated, despite counsel for more pilot work; less than 10 succeeded. Some reasons for failure are that the randomization is contaminated and that the treatments are not delivered as advertised.

Other problems with statistical evaluation are that treatments are not "fixed" in the same sense that treatments are fixed in the chemical sciences or engineering, and that response variables are measured poorly or are irrelevant to the treatment. Boruch also pointed out that "many educational and social problems are chronic. They are resolved in small steps over a long period despite innovative social programs dedicated to their elimination, and despite inspirational rhetoric."

Boruch observed that there are certain forms of corruption of measurement "that have not been well articulated in the orthodox literature on randomized experiment." He maintained that these problems deserve much more attention from the statistics community. But in my opinion there is no statistical *formula* that can deal with the corruption of measurement. Scientists simply have to use their subject-matter knowledge to develop better instruments and to collect better data.

Boruch also recommended estimates based on a combination of a few expensive randomized tests and a large group of cheaper nonrandomized tests. But there are some general problems in this direction (Boruch, 1982):

It is not clear whether and how to design for sequential trials in which randomized tests alternate with nonrandom tests.

It is not clear how one ought to design simultaneous randomized and nonrandomized trials in anticipation of internal analyses that must ignore the randomization category at least partly but that are also used to inform policy.

Nor is it clear how, when one is confronted with a set of ostensibly independent randomized and nonrandomized field tests, to decide whether combining estimates of effect is warranted or how to combine them.

Boruch finally appealed to the statistical community for vigorous participa-tion in solving these problems. This appeal, albeit published in a popular jour-nal, *The American Statistician*, has not been well received. The reasons are simple: Boruch was asking for the impossible. Statistical formulas are *blind* to whether a test was randomized or not randomized. It is a scientist's job to ran-domize the treatment, and then to use *professional judgment* to draw infer-ences from a set of randomized and nonrandomized tests. This professional judgment is largely content-specific, and statistical formulas in general will not help. A rare exception is the Bayes theorem. But first, scientists have to devise a reliable measure of the quality of nonrandomized tests. Otherwise the Bayesian approach will produce more uncertainty in the study.

In summary, there is no easy way to conduct a credible randomized test in an educational setting; there is no formula to measure the quality of nonrandomized tests; and there is no sensible procedure to combine the information obtained separately from randomized and nonrandomized tests. In other words, it is a jungle out there. But if you are in a Tarzan mold, then this jungle may well be the closest thing to heaven for you. This is because experimentation is not the only way that knowledge is acquired. For example, Darwin did not conduct randomized comparison in order to develop his theory of evolution—he would *never* have succeeded if he had.

The field of evaluation research was dominated by statistical methodology prior to the late 1970s. Now there are voices in opposition, and some researchers are advocating a qualitative approach with admirable vehemence (see, e.g., Patton, 1980). This new perspective uses subject-matter knowledge, expert judgment, intellectual debates, and plain descriptive statistics. In sum, one has to use his eyes and mind; textbook statistical formulas are far less useful tools.

II. MODELING INTERACTION EFFECTS: A CASE STUDY FROM THE SOCIAL- BEHAVIORAL SCIENCES

Throughout this book we have been very critical of "quantitative analysis" in the social, economic, and behavioral sciences. However, this criticism does not mean that no research findings in these disciplines deserve respect. On the contrary, in these disciplines many *qualitative* analyses are highly admirable, and genuinely enhance our understanding of our society and ourselves.

A famous example is Sandra Bem's theory regarding masculinity and femininity. The theory was ground-breaking and touched off an avalanche of research publications. The theory even made its way into a Presidential Address of the American Statistical Association (Kruskal, 1988, *JASA*) on "The Causal Assumption of Independence."

The traditional (and still popular) views of masculinity and femininity are very narrow. For example, according to the Thorndike and Barnhart Dictionary, "masculine" simply means "manly," or having characteristics of that sort. For another example, several recent studies showed that psychotherapy and psychiatric diagnosis are often used to enforce conventional standards of masculinity and femininity (*The New York Times*, April 10, 1990). In such cases, stereotypes about men and women subtly distort diagnosis and treatment: men and women who deviate from traditional sex-roles are often seen as suffering a mental illness.

Bem (1974) set out to challenge the traditional perspective about sex-roles in our society. To do so, she first distributed questionnaire items to a group of

Stanford University students (n = 100). The responses of these students
identified "masculine" as competitive, assertive, analytical, ambitious, willing
to take a stand, etc., and "feminine" as gentle, understanding, affectionate,
having concern for others, etc.

Bem's genius was that she recognized that psychological masculinity and
femininity are not opposite poles of a single dimension:

Rather, they are two *orthogonal* personality dimensions on which individuals
might vary independently:

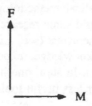

According to Bem (1974), (i) it is conceivable that an individual may be
psychologically both masculine and feminine; and the existence of such in the
same person is then defined as androgyny (andro—king; gyny—queen); (ii)
strongly sex-typed individuals might be seriously limited in the range of
behaviors available to them as they move from situation to situation, and therefore it is hypothesized that androgynous individuals are mentally healthier and
socially more effective.

The concept of psychological androgyny was completely new in 1974, and
soon became a major focus in social-psychological research. A burgeoning
literature grows to assess the above two hypotheses and to find associations
between androgyny and a wide range of variables (z): mental health, personal
adjustment, self- esteem, achievement, autonomy, and so on. (A computerized
reference search by Lipta (1985) yielded 432 research studies employing
Bem's sex-role inventory.)

A.   The Quantification of Psychological Androgyny

A predominant fact in "modern" social-behavioral sciences is that in order to
gain scholarly recognition, your theory has to be "justified" by statistical tests.
There are plenty of good reasons why this should be the case, but the problem
is that many quantitative analyses in behavioral research are misleading and
lacking in intellectual depth. In this section we will examine some of the statistical analyses carried out by Bem and her followers in their inquiries into
psychological androgyny.

First, from students' responses on a pool of 200 items, Bem used a statistical test to identify 20 items for M (masculinity) and 20 for F (femininity) scales. Each item ranged from 1 (never or almost never true) to 7 (always or almost always true). The resulting scales serve the foundation of almost all later androgyny research. A critique of this foundation is given in Section II.F. of this chapter.

The central idea of Bem's theory is that many individuals might be "androgynous." That is, they might be both masculine and feminine, both assertive and yielding, both instrumental and expressive—depending on the situational appropriateness of these various behaviors. Also, they are mentally healthier and socially more effective than other people.

To test this hypothesis, it is desirable to have a measure of androgyny. To this end, Bem (1974) proposed the following t-ratio as a measure of psychological androgyny:

$$t = (M - F)/SE(M - F), \tag{1}$$

where the numerator, $M - F$, reflects the balance concept of psychological androgyny, while $SE(M - F)$ is the standard error of $(M - F)$ of the population under study. Subjects are classified as sex-typed if the androgyny t-ratio reaches statistical significance ($|t| > 2.025$, d.f. $= 38$, $p < .05$), and they are classified as androgynous if $|t| < 1$. (Bem, 1974, p. 161).

According to formula (1), a balanced individual with $(M,F) = (4,4)$ is more androgynous than a sex-typed person with $(M,F) = (7,1)$; thus the t-ratio does capture the balance concept of Bem's theory. But there is a major flaw: Subjects with low scores for both M and F (which are less socially desirable) are erroneously classified as androgynous. For instance, a person with $(M,F) = (1,1)$ would likely be perceived as a wimp, but he has the same androgyny score as people with highly desirable $(M,F) = (7,7)$. This internal inconsistency finally led to the abandonment of the t-ratio as a measure of androgyny.

Following the t-ratio, a median-split technique was used for classification (Spence et al., 1975; Bem, 1977):

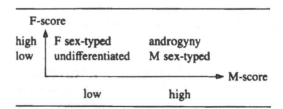

This simple strategy makes a lot of sense and is very popular among researchers who use analysis of variance (ANOVA) to assess the relationships between androgyny and a host of dependent variables (y).

The median-split method has, nevertheless, been viewed as unsatisfactory because of the loss of information in using only four categories to classify the whole population. As a result, different models have been proposed to construct formulas for androgyny on a continuous scale.

Spence et al. (1975) proposed the following additive model of androgyny:

$$A = M + F. \tag{2}$$

The problem with the additive model is that it does not take into account the possible interaction effect of M and F. Also, a balanced individual with (M,F) = (4,4) is as androgynous as a sex-typed person with (M,F) = (7,1). Thus formula (2) is a serious departure from Bem's original idea.

Lubinski et al. (1981, 1983) argued for the inclusion of an interaction term in the model, on the ground that the androgyny concept theorized by Bem was interactive in nature. Moreover, to achieve the status of an empirically useful concept, androgyny must have unique properties beyond those possessed by masculinity and femininity.

B. Models for Interaction Effect: A Review

The simple multiplication of M and F

$$A = MF \tag{3}$$

represents the symbolic form of Bem's balance concept (see Hall and Taylor, 1985, p. 431), and was one of the early models for the interaction effect. Lubinski et al. (1981, 1983) advocated the following regression model

$$z = f(M,F) = a + bM + cF + dMF. \tag{4}$$

The model, which combines additive and multiplicative terms, was used to predict z (a dependent variable) by the main effects (M,F) and the M × F interaction. This model is thus far the most popular model in the androgyny literature. A critical review of the model is rather involved and is presented in Section II.C. of this chapter.

Bryan et al. (1981) proposed the following geometric-mean model:

$$A = (MF)^{1/2}. \tag{5}$$

An interesting feature of this model is that the formula yields near-normal distributions (although theoretically it may produce imaginary numbers, assuming that M and F are independent and normally distributed). This model can be viewed as a special case of a family of models in formula (11), Section II.E. of this chapter.

Heilbrum and Schwartz (1976, 1979, and 1982) published at least three articles on a hybrid model

$$A = (M + F) \times |(M - F)| \tag{6}$$

which incorporated the balance concept $|M - F|$ into the additive model $(M + F)$. But this formula is equivalent to $|M^2 - F^2|$, and suffers the same problem as $|M - F|$ or the t-ratio in formula (1).

Karlin (1979) suggested $(M + F) + |M - F|$ as another measure of androgyny. The motivation behind this model appears similar to that underlying formula (6). The Karlin model is a special case of (7) below, which was proposed by Hall and Taylor (1985).

Hall and Taylor (1985) discussed strengths and weaknesses of formulas (2)-(4), and concluded with deep disappointment in those models. They further explored the feasibility of using the following model:

$$z = f(M,F) = a + bM + cF + d|M - F| \tag{7}$$

which differs from the regression model [formula (4)] in the last term. It was found that this formula would provide a simple operational definition of the balance concept only under highly restricted conditions, and thus failed to serve as a general formula in a wide range of situations.

For future androgyny research, Hall and Taylor concluded that "an accurate interpretation of interaction effects will be crucial." We accept this challenge, and will present new models after discussing the regression model and the shape analysis of existing models.

C. A Misconception of Regression Model

A widespread confusion in regression analysis arises from the interpretation of the cross term MF. This misconception has deep roots in the statistical literature. For instance, Nelder (*Statistical Science*, 1986) attempted to use the cross term to model the interaction effect.

It should be noted that this cross term is useful for *testing* the existence of interaction effects. However, the equation

$$z = f(x,y) = a + bx + cy + dxy \tag{8}$$

itself does not promise to give a good *fit* to the interaction. The reason is that the surface of equation (8) is a horse saddle (See Fig. 2).

In 3-dimensional space, the shape of the function $z = f(x,y)$ can be very complicated. For instance, if

$$z = \sin[(x^2 + y^2)^{1/2}] \qquad -5 < x < 5, \qquad -5 < y < 5,$$

then the shape of the function is a cowboy hat (see Fig. 3). On a farm, it would be a joke to use a cowboy hat as a horse saddle; but in statistical modeling, it is done all the time.

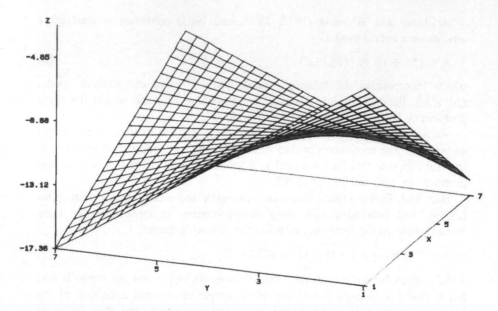

Figure 2   Graphical display of the regression model $Z = a - 2.527 * X - 2.707 * Y + .588 * X * Y$.

In many branches of natural science, sophisticated differential equations are often used to model interaction effects. Further, the modeling is mostly content-specific, and there simply does not exist a catch-all formula that would be applicable to all disciplines. In summary, one has to be careful not to confuse "testing" with "modeling." Unfortunately, the confusion has diffused from statistical literature to many branches of empirical science. Let us consider an example from the androgyny literature.

Let Z_1 be a measure of status concern and Z_2 be a measure of impulsivity. Anderson (1986, p. 273) found that (for males)

$$Z_1 = a_1 - 2.527M - 2.707F + .588MF, \tag{9}$$

and

$$Z_2 = a_2 + 3.372M + 3.793F - .729MF, \tag{10}$$

where a_1 and a_2 are constants which were not reported in Anderson's paper. Anderson (p. 274) interpreted the *main effects* M and F as follows: "Both high M and high F are associated with lower status concern and greater impulsivity." The *interaction effects* were interpreted: "The two interactions indicate that among men who are high in masculinity, femininity increases status concern (Z_1) and lowers impulsivity (Z_2). On the other hand, among men who

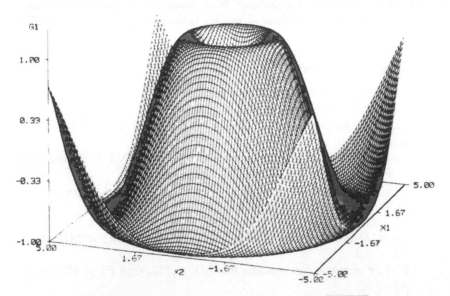

Figure 3   Graphical   display   of   the   equation $Z = \sin(\sqrt{X^2 + Y^2})$ $-5 < X < 5$, $-5 < Y < 5$.

are average to low in masculinity, femininity lowers status concern and increases impulsivity. Both of these are opposite to the androgyny prediction."

The above interpretations of main effects and interaction effects are very common and confusing. The surface of equation (9) is indeed the previous saddle, with saddle point near (4,4). That is, an individual with (M,F) = (1,1) is more androgynous than a person with (M,F) = (4,4). This does not appear to make sense. In addition, the R-square (the so-called percent of explained variability) of equation (9) is only 1.8% (and 3.1% for equation (10)). This low value of R-square indicates that neither equation (9) nor (10) really explains much of the phenomenon.

D.   Shape Analysis and the Rationale of New Models

It is a common belief that in the social-behavioral sciences nonlinear models amount to overkill. But in androgyny research, qualitative analysis indicates clearly that the phenomenon under study is nonlinear. If quantitative psychologists want to progress, an operational formula of a nonlinear nature is inevitable.

In this section we examine some mathematical properties of the differently defined A's which have been implicitly assumed in the androgyny literature. It is important to point out that some of these properties *contradict* each other and that a careful analysis of these properties is the crux of the development of new models.

In the following discussions of P1–P4, let (M1,F1) and (M2,F2) be the M and F scores of any two individuals.

1. P1. *Diagonal monotonicity*: Assume that $(M1 - F1) = (M2 - F2)$; then $M1 < M2$ implies $A1 < A2$.

2. P2. *The balance property*: If $|M1 - F1| < |M2 - F2|$, then $A1 > A2$.

   P2-S. *The semi-balance property*: If $M1 + F1 = M2 + F2$ and $|M1 - F1| < |M2 - F2|$, then $A1 > A2$.

   P2-Q. *The quasi-balance property*: $f(M,M) > f(M - u,F)$, for all $1 < M < 7$ and for all F, while $0 < u < 1$.

3. P3. *Additivity*: If $M1 + F1 < M2 + F2$, then $A1 < A2$.

   P3-M. *M-monotonicity*: Assume that $F1 = F2$; then $M1 < M2$ implies $A1 < A2$.

   P3-F. *F-monotonicity*: Assume that $M1 = M2$; then $F1 < F2$ implies $A1 < A2$.

4. P4. *Normality*: If M and F are normally distributed, then the distribution of A is normal or near normal.

Discussions of P1 – P4

P1 (diagonal monotonicity). Following the hypothesis implicitly assumed in the median-split technique, P1 represents a more precise mathematical formulation of this hypothesis. Specifically, individuals located higher on the main diagonal of the MF-space are expected to have a higher A score than those located lower. The same property applies to the lines parallel to the main diagonal. It is obvious that formulas (1), (6), (7), and (8) do not satisfy P1.

P2 (the balance property). This property is central to the original androgyny concept (Bem, 1974). However, if a model satisfies P2, e.g., formulas (1) and (6), then it is hard to satisfy other desirable properties. For this reason, one has to look for less stringent conditions, e.g., P2-S and P2-Q.

P2-S (the semi-balance property). This property says that on a line perpendicular to the diagonal, the smaller the difference $|M - F|$ is, the larger A should be. This property is relatively easy to achieve, in the sense that most models considered in this chapter satisfy P2-S.

P2-Q (the quasi-balance property). Note that P2-Q is a special case of P2. However, P2-Q is not shared by many popular models [e.g., formulas (2), (3), (4), and (5)].

P3 (the additive property). The rationale behind P3 is that high M and high F are socially desirable. However, mathematically P3 *cannot coexist* with P2. Furthermore, they contradict Bem's original idea that extreme degrees of either M or F can become negative and even destructive. Indeed, many popular models [e.g., formulas (2), (3), (4), and (5)] are modeling P3, not P2.

Note that P3-M and P3-F are special cases of P3, but P3-M and P3-F do not imply P3. For example, if $f(M,F) = (MF)^{1/2}$, then f satisfies P3-M and P3-F, but $f(1,7) < f(2,4)$ and $f(2,7) < f(4,4)$, violating P3.

P4 (the normality property). It is generally assumed in psychological literature that certain measures of psychological functioning are normally distributed. Since all theories of androgyny assume close relationships between A and z variables (e.g., mental health, flexibility), it follows that A is expected to be normally distributed as well. It should be further pointed out that empirical studies have shown the normality of M and F scores (this is expected because each of these scores represents an average of ratings on a 7-point scale for 20 items). A formidable challenge is to propose a formula which would satisfy P4 (the normality assumption) and at the same time catch the intrinsic complexity of the interaction effect of M and F. For example, if M and F are i. i. d. Normal(4,1), and

$$A = 1/[(M - F)^2 + 1]^{1/2},$$

then A certainly satifies P2 (the balance property); but the distribution of A looks like Fig. 4, which is far from being normal and may create difficulty in relating A to certain y variables.

E. New Models for the Interaction Effect

We now present a formula to model the interaction effect of M and F:

$$A = rMF/[(M - F)^2 + eMF]^{1/2} - a|M - F|; \qquad e > 0; \qquad ra > 0 \quad (11)$$

The rationale of this model is rather complicated and will be made available upon request. The advantage of this model is that it satifies P1 and almost P2 and P4.

THEOREM If

$$A = f(M,F) = MF/[(M - F)^2 + eMF]^{1/2} \qquad (12)$$

then the following results hold.

1. A satisfies P1 (the diagonal monotonicity property).
2. A satisfies P2-S (the semi-balance property).
3. If $e < u$, then A satisfies P2-Q (the quasi-balance property).

Proof. Available upon request.

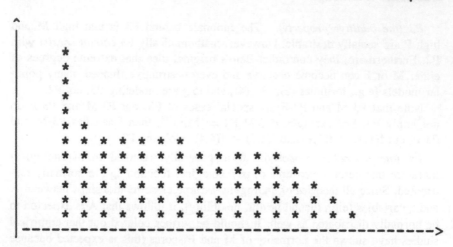

Figure 4 Frequency distribution of the random variable $1/\sqrt{(M - F)^2 + 1}$, where M and F are i.i.d. Normal (4,1).

Remark 1. In practice one may choose e to be less than or equal to .1. For this choice of e, A almost satisfies P2, a major mission of the androgyny modeling. If e = .1, some special values of A in (12) can be calculated as in the table:

F							
7	1.2	2.7	4.9	8.2	912.8	18.4	22.1
6	1.2	2.9	5.5	9.5	15.0	19.0	18.4
5	1.2	3.2	6.4	11.6	15.8	15.0	12.8
4	1.3	3.7	8.1	12.7	11.6	9.5	8.2
3	1.5	4.7	9.5	8.1	6.4	5.5	4.9
2	1.8	6.3	4.7	3.7	3.2	2.9	2.7
1	3.2	1.8	1.5	1.3	1.2	1.2	1.2
	1	2	3	4	5	6	7 M

The geometric shape of the table is shown in Fig. 5.

Remark 2 (the normality assumption). If X and Y are independent normal with zero means and standard deviations σ_1 and σ_2, then it can be proved (Feller, 1971, Vol. 2, p. 64) that

$$XY/(X^2 + Y^2)^{1/2} \qquad (14)$$

is normal with standard deviation σ_3 such that $1/\sigma_3 = 1/\sigma_1 + 1/\sigma_2$. For this

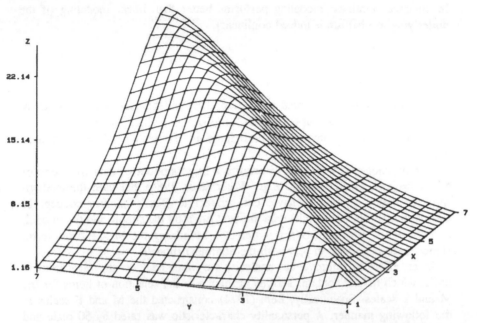

Figure 5 Graphical display of the equation $A = MF/\sqrt{(M - F)^2 + .1 * M * F}$

case study, formula (12) yields only a near-normal distribution. But this is close enough for most applications.

Remark 3. The regression model, equation (12) or (13), is a special case of the Taylor expansion. If f(M,F) is smooth, mathematically one can approximate a nonlinear function by higher order Taylor polynomials (in a neighborhood of a given point). A simulation study (see the following table) shows that a portmanteau model utilizing Taylor polynomials can be far from satisfactory (just compare the SSE's of the fitted models with the expected SSE).

Eq.	Constant	M	F	MF	M^2	F^2	SSE
1				2.5			305,944
2	148	−33.7	−34.6	10.4			232,507
3		1.3	.7	2.0			303,870
4		3.4	6.9	11.0	−4.9	−5.4	123,075
						Expected SSE	= 1,000

The simulation model is as in (11): $r = 3$, $e = .03$, $a = 0$, $y = A + Z$, Z is standard normal, and $n = 1000$. The geometric-mean model has been used to fit the data, and the model yields a SSE of 268,863. A third-order Taylor expansion has also been fitted to the data, and yields a SSE of 130,775—a very poor fit.

In contrast, nonlinear modeling performs better than linear modeling (if the underlying mechanism is indeed nonlinear).

F. When Is a Science Not a Science: Flat-Earth Theory and Stone-Age Measurements

The nonlinear model presented in Section II.E will be useless for androgyny research, if it is not supported by empirical results. For this reason a psychometrician was invited to participate in the testing of the nonlinear model.

The psychologist was swift to acquire a huge amount of data from one of his colleagues. Earlier results (e.g., Watermann, 1984) that used the median-split technique claimed to have lent support to Bem's balance concept of androgyny. But in our test the data are best fitted by the simple additive model, $z = aM + bF$. No interaction was detected by adding the $M \times F$ cross-term. Hence there is no need to play with the nonlinear models.

In an effort to improve the analysis, we went back to the measurement itself, which indeed contains serious problems in the selection of items for the M and F scales. Specifically, Bem (1974) constructed the M and F scales in the following manner: A personality characteristic was rated by 50 male and 50 female students; it was qualified as masculine "if it was independently judged by both males and females in both samples to be significantly more desirable for a man than for a woman ($p < .05$)."

This process represents a typical confusion between practical significance and statistical significance. As a result, out of 20 items for femininity scores, some (e.g., gullible and flatterable) are not desirable, at least not in American society. Also, both M and F scales contain, respectively, "masculine" and "feminine" items, which measure biological androgyny, not psychological androgyny. Because of such internal contradictions, Bem (1979) developed a short BSRI (Bem Sex-Role Inventory) containing exactly half the number of the original items.

We tried to test the interaction effect with the short BSRI. But there is a problem: In the data set we borrowed, the 20 M-items had been averaged out to the M score, and the original data had been thrown away. We looked at that huge amount of data and were deeply disappointed.

The same situation occurred when we requested raw data that were used in Anderson (1986): The scores for separate items are not available; they have been added together. It appears that researchers and journal editors in psychology are more familiar with the original BSRI than the short, improved BSRI. It is often said that science is self-correcting. This is not quite true, at least not in behavioral sciences. In this case study, researchers keep pumping out publications without a proper concern for the accuracy of the instrument used in data collection.

Once I asked my students the following question: "How did a man measure a piece of land 5,000 years ago?" They answered: "He walked and counted the feet." I further asked: "How does an engineer measure the height of a mountain today?" They answered: "He uses laser beams." In comparison to engineering measurements, most modern-day psychometric instruments are still in the stone age.

It is important to note that there is nothing to laugh about at these stone-age instruments—they are quite natural and understandable to a young discipline. At least these instruments are better than no instruments at all. What is troubling is that journal editors and referees are so careless about the instruments used in scientific investigation. Even if they do not keep up with the literature on the short BSRI, simply by common sense one can smell the rot in the original instrument.

Richard Feynman (a world-renowned physicist) once said in an interview on the television program NOVA, "Social science is an example of science that is not a science. They don't do scientific ... they follow the form ... or you gather data, you do so and so and so forth" [sic]. From my experience with the androgyny research, it appears that Feynman is not so far off. Maybe "science" is too generous a word for this type of research.

Moving back to the issue of measurement, one can easily detect several drawbacks in the short BSRI: (1) The criterion for item selection is based on the loadings of factor analysis and item-total correlations. These statistical tools somehow took the place of a scientist's instinct and subject-matter knowledge. (2) By common sense there appear to be several redundant items: for example, "defends own beliefs" and "willing to take a stand" in the M-scale, "sympathetic" and "compassionate" in the F-scale. For each of those pairs, Roget's Thesaurus classifies them as having the same or very similar meanings. (3) Bem's adherence to having exactly 10 M-items and 10 F-items might have excluded other good items from the inventory. "Ten" may be a perfect number, but 10 items do not necessarily add up to a perfect scale. In fact, unequal numbers of M and F items can easily be weighted to produce a better instrument.

My co-worker later borrowed more data sets from other sources. But the y-variables in these data sets are remotely correlated with the M-score and in many cases independent of the F-score. Therefore no interaction effect was detected.

Because of the overwhelming "evidence" that the interaction is not detectable, my co-worker felt that the balance concept of androgyny is not supported by data, and that the additive model apparently represents the phenomenon better. He maintained that we ought to trust the data, and that this is what empiricism is all about.

I am a statistician, but I am not an empiricist who trusts everything that data have shown. Let us summarize the reasons why the interaction effect should be there: (1) If we accept the additive model, $A = M + F$, then $A(4,4) = A(7,1)$. That is, a balanced individual with $(M,F) = (4,4)$ is the same as a highly sex-typed person with $(M,F) = (7,1)$. This notion simply does not make any sense. (2) Most psychometric measurements contain an enormous amount of noise. Therefore one cannot take the results seriously when this kind of measurement declares that the test of interaction is not statistically significant. (3) Most psychometric studies are based on samples of college students, or simply the students in a Psychology 101 class. These students usually have very similar backgrounds, and constitute only a tiny portion of a huge population. It is then predictable that this type of data will produce an additive model which has a flat surface. As an analogy, suppose that a million people measure their backyards, and none can detect the curvature of the Earth. Can we take these one-million empirical studies as "evidence" and claim that the Earth is flat?

Unfortunately, these arguments did not convince my co-worker. He kept citing leading researchers (e.g., Lubinski et al., 1981, 1983) to buttress his conclusion that there is little empirical support for the interaction of M and F.[9]

G. Reliability and Validity of Scientific Measurements (Part I)

In order to test better the existence of an interaction effect, I told my co-worker that we could not keep going after somebody's data sets that were aimed at other research foci. Instead we had to find a good instrument, and collect data relevant to our own project. A few weeks later I received a thick package and a message from my co-worker that we were ready to collect our own data (with the assistance of two psychology majors).

The instrument that would be used is called POI (Personal Orientation Inventory). According to Shostrom (1980) there were more than 800 references related to POI. This whopping list of publications is very impressive. But a bitter lesson I have learned is that the quantity of publications cannot by itself be taken as the foundation for further research. One has to know at least what the instrument is trying to measure.

The inventory was created to gauge an interesting concept called "self-actualization." Among scientists or academicians, many have gone through a process of self-development through their professional activity (no matter how narrow or esoteric it is). If well-explored, the concept of "self-actualization" may reveal a fascinating world of our mental activities.

According to Shostrom (1980), a self-actualizing individual is seen as "developing and utilizing all of his unique capabilities, or potentialities." Such an individual is "free of the inhibitions and emotional turmoil of those less self-actualizing."[10]

The complete POI has 150 items (printed in 6 pages) for two basic scales and 10 subscales, each of which measures a conceptually important element of self-actualization. Some of the subscales are very interesting. For example, "spontaneity" measures freedom to react spontaneously or to be oneself; "existentiality" measures ability to react situationally without rigid adherence to principles. In other words, people who get low scores on the Ex-scale (existentiality) tend to hold values so rigidly that they may become compulsive or dogmatic (Shostrom, 1980, p. 17).

These descriptions of "existentiality" make a lot of sense. But the quantification of the descriptions is, ironically, extremely rigid. Here are five items out of the first seven in the Ex-scale:

a. I am concerned with self-improvement *at all times*.
b. I am not concerned with self-improvement *at all times*.

a. I feel that I *must strive for perfection in everything that I undertake*.
b. I do not feel that I *must* strive for perfection in everything that I undertake.

a. I feel I *must always* tell the truth.
b. I do not *always* tell the truth.

a. I am bound by the principle of fairness.
b. I am not *absolutely* bound by the principle of fairness.

a. When a friend does me a favor, I feel that I *must* return it.
b. When a friend does me a favor, I do not feel that I *must* return it.
[emphases supplied]

Students are expected to select either (a) or (b) in each item. With wording so rigid and absolute, it is hard to believe that these items will yield any useful information. A scale that ranges from 0–4 or 1–5 might tell us more about an individual. But my co-worker was against changing the scale of the measurement. The reason is that it would affect the reliability and validity of the instrument.

This so-called "reliability" is used in the construction of almost all psychological instruments. For the case of POI, the reliability was calculated on the basis of a sample of 48 undergraduate college students. The inventory was administered twice, a week apart. Mathematically, let $X_1, \ldots, X_n$ be the scores of the first week, $Y_1, \ldots, Y_n$ be the scores of the second week, and the Pearson correlation coefficient of X and Y is called the "test-retest reliability" of a psychometric instrument. For POI, this correlation coefficient is about .75. That is, the R-square of Y versus X is about 56%. In plain English, you give the same group of students the same test a second time. One week later the test retains only 56% of the original variability.

According to the manual (Shostrom, 1980), POI is a useful tool for practicing counselors in psychological therapy; POI is also related to "actualizing therapy: foundations for a *scientific ethic*" [emphasis supplied]. This scientific foundation and the related 800 publications, in view of the POI's reliability, are things the scientific community cannot be proud of.

POI is also depicted as a "standardized test" containing "objective words." However, its reliability indicates that the instrument is quite arbitrary, and that it should be used for exploratory purposes, instead of as a rigid "standardized test." Finally, a better way to do *exploratory analysis* is to change the original {0,1}-scale into {0,1,2,3,4}, 1-5, or 1-10. But my co-worker was very reluctant to temper with the scale. After a long debate, our collaboration dissolved, and the nonlinear model of androgyny remained untested.

H. Reliability and Validity of Scientific Measures   (Part II): A Religious Icon in Behavioral Research

In this section we will discuss a case study in which psychometric reliability and validity undermined an otherwise illuminating study.

A major task of our liberal-arts education is to teach our students how to think independently, how to establish their own goals for their lives, and how to make judgments that are best for themselves. If our citizens cannot think independently and rationally, as some have argued, the foundation of our democratic society will weaken. In professional disciplines, if our students are trained to follow authorities blindly, then we will drain their creativity and reduce their opportunities for improvement.

For these reasons, many studies are aimed at *finding out* how autonomous our students really are. In one instance, an investigator attempted to use the BSRI and another psychometric instrument for an autonomy study. The investigator also drew up some questions that were thought to be pertinent to the study. The scale for each question ranged from zero to four. This scale makes sense, because it is parallel to students' GPA and hence is likely to reflect students' true feelings about each question.

Unfortunately, the investigator asked the opinion of a quantitative psychologist (an *official statistical consultant* in an academic institution). The consultant told the investigator that BSRI is scaled 1-7, and thus the scaling of the last 40 items in the questionnaire had to be changed from 0-4 to 1-7.

The reason for my dissent is that the single most important thing in scientific investigation is good measurement. If you don't have good measurements, then no statistical mumbo-jumbo can save you. In this case, the questionnaire will be administered to students in numerous classes of various (absent-minded) professors who are willing to help in collecting data. The students are expected to answer the questions scaled 0-4 on the first two pages and then change to 1–7 on another 40 questions. What will happen if they do not notice the change of the scales?

In my own experience, I've found that students are very careless about filling out questionnaires. And it is not a good thing to confuse students with two different scales in the same questionnaire. Also, garbage in the original BSRI should be trimmed away—a long questionnaire only creates more noise, not more information.[11]

The psychologist was not moved by my concern regarding measurement. He insisted that these psychometric instruments are well-established and that if we changed the scale then we would destroy the reliability and validity of the instrument. To him the reliability and validity measures of psychometric instruments are something from the Holy Bible that nobody should alter— although he had worked on this instrument and knew how poor it was.

The test-retest reliability of BSRI was similar to that of POI. Based on the Pearson correlation coefficients, Bem (1974) concluded: "All four scores [in BSRI] proved to be *highly reliable* over the four-week interval (Masculinity $r = .90$; Femininity $r = .90$; Androgyny $r = .93$; Social Desirability $r = .89$)" [emphasis supplied].

Despite Bem's claim that BSRI is highly reliable and "internally consistent" (p. 160), some intelligent psychometricians (e.g., Pedhazur, 1976) pointed out serious problems in the BSRI: The whole Androgyny score and many items in the Masculinity and Femininity scores simply do not make any sense. In fact, Bem herself (1979) abandoned the Androgyny score completely, even though it has an extremely high reliability ($r = .93$). She also reduced the items in both Masculinity ($r = .90$) and Femininity ($r = .90$) to half. Thus it is difficult to understand why the 1974 instrument is treated like the Ten Commandments that one has to obey without any reservation.

As a comparison, let's see how reliability is established in engineering. For example, how do we know that a Toyota is reliable? The Consumers Union and Toyota Inc. keep extensive repair records of the automobiles, and the engineers in charge constantly work on the reduction of the defects in their automobiles. They do not merely calculate product-moment correlations and then claim that the vehicle is highly reliable.

We also know that SAT scores are reliable to a certain extent. How is this achieved? A group of researchers at ETS (Educational Testing Service) in Princeton make life-time careers of debugging the test items. These people have to guard the integrity of the SAT against unfair test-takers and frequent charges of sex-bias in the tests. They do not achieve this by calculating a product-moment correlation coefficient in 1974 and then using the test until 1999.

A defense that one should stick with the original scale may go like this: If every investigator used the scales that he or she liked, then nobody would be able to do a comparative study. This argument is in general acceptable; but it does not apply here. First, if one uses the scale 0–4, one can always use a simple formula $(1 + 1.5X)$ to convert to the original scale 1–7. This conversion may distort the instrument in some subtle ways. But this instrument is already 15 years old, and nothing can be more important than reducing students' confusion in using two scales in one questionnaire.[12]

Bem's 1974 paper was a breakthrough for its insight (i.e., qualitative analysis); but the quantitative analysis in the paper was far less successful. Researchers who want to use BSRI, POI, or similar instruments have to be aware that the so-called test-retest reliability is at best a benchmark for cross-validation. The investigators still need their eyes and minds when they use these psychometric instruments.

NOTES

1. I was given only one weekend to review the thick proposal.
2. That was stupid, but it was not the only stupid thing I have done in my life.
3. Similar mentality of this two-sample comparison can be found in a popular textbook in the design of experiments (Hicks, 1982, p. 30, Example 2). The Hicks study compared a group of 10 students before and after a six-week instruction $(t = 3.56, P < .005)$, and concluded that "there has been an improvement" in achievement. But important information was mostly missing in Hicks' report. For instance, the differences in the test scores averaged only 2.5 (out of 100, maybe). Even if the difference was practically significant, it might be due to the similarity of the tests given before and after the instruction.
4. As a recent example, mathematician Serge Lang waged a war against the nomination of a Harvard political scientist to the American Academy of Science. The Harvard scholar uses mathematical equations heavily in his publications in political science. Fortunately, Lang won the battle.
5. If we have to bring down the whole house of educational statistics, so be it. In a recent *Journal of Educational Statistics* (1987, Vol. 12, No. 2), David Freedman of University of California at Berkeley launched a devastating attack on the use of multiple regression in social-science research. Freedman's article was commented on by 11 eminent scholars. Although Freedman won the argument, I believe the practice remains "business as usual."
6. A major problem with external evaluation is that the budget was not built in the original proposal. I suggested to the Director that they should purchase site licensing of the software (instead of buying 100 copies for all the students). This may save them $1,000 so that they can have money to recruit other instructors to repeat the experiment in Spring 1990.

     The director mentioned that there was another problem: he was bound to the statements in the proposal and hence must include the control groups in the experiment. Well, it is a conventional wisdom that you should consult a statistician before you conduct an experiment. We now learned a further lesson from this case study: consult your statistician before you write up your proposal. Also, give your statistician plenty of time. One weekend is simply not enough.
7. Later developments: After reading all of the above analysis, the director agreed to

reduce the three control groups to one in the Fall semester. But he asked a curious question: Why is the two-sample comparison not scientific?

I don't know if he was confused, or I was confused, or both of us were confused. I hope that somewhere Neyman and Pearson had said that "the two-sample test may not be appropriate in educational comparisons," so I can quote authority and save all the trouble.

Ten days later, the director said they might increase the number of control groups to two (in the Fall semester). I was too exhausted to ask for the reasons. Like a medical doctor, I have given you my medicine. If you don't want to take it, that's your problem, not mine.

8. Heid interviewed 20 students, 15 from the experimental classes and 5 from the comparison class. This random sample is rather *unfortunate*. For a situation like this, a nonrandom sample (with careful matching) would be better than a random sample. At least Heid should have interviewed a more nearly equal number of students from the comparison class.

9. Many researchers found that M and y (mental health, etc.) are highly correlated, while the correlations between F and y variables are moderate or negligible (using the original BSRI). This is because some undesirable items (gullible, flatterable, etc.) were hidden in the F-score. As a result, negligible correlation between F and certain y variables undermines the interactive concept of androgyny and leads many researchers to believe the additive model.

10. This description of a self-actualizing person does not seem consistent. We know many geniuses are free of inhibitions (e.g., Beethoven, Van Gogh, etc.), but they are famous for their severe emotional turmoil. Maybe these geniuses are statistical outliers beyond current quantitative psychology.

11. In some questionnaires, redundant questions are used for cross-validation. But this is a different story.

12. If one insists on using the original scales (0–4 and 1–7), there are many ways to include cautious notes to reduce the confusion. But such precautions never entered the debate.

REFERENCES

Anderson, K.L. (1986). "Androgyny, Flexibility, and Individualism," *J. of Personality Assessment*, Vol. 50, 265–278.

Bem, S.L. (1974). "The Measurement of Psychological Androgyny," *J. of Consulting and Clinical Psychology*, Vol. 42, 155–162.

Bem, S.L. (1977). "On the Utility of Alternate Procedures for Assessing Psychological Androgyny," *J. of Consulting and Clinical Psychology*, Vol. 45, 196–205.

Bem, S.L. (1979). "Theory and Measurement of Androgyny: A reply to the Pedhazur-Tetenbaum and Locksley-Colten critiques," *J. of Personality and Social Psychology*, Vol. 37, 1047–1054.

Boruch, R.F. (1982). Experimental Tests in Education: Recommendations from the Holtzman Report. *American Statistician*, Vol. 36, No. 1, 1–14.

Bryan, L., Coleman, M., and Ganong, L. (1981). "Geometric Mean as a Continuous Measure of Androgyny," *Psychological Reports*, Vol. 48, 691–694.

Deming, W.E. (1982). *Quality, Productivity, and Competitive Position*. Massachusetts Institute of Technology, Center for Advanced Engineering Study.

Feller, W. (1971). *An introduction to Probability Theory and Its Applications*, Vol. II, second edition. Wiley, New York.

Freedman, D. A. (1987). As Others See Us: A Case Study in Path Analysis. *J. of Educational Statistics*, Vol. 12, No. 2, 101–128, 206–223; with commentaries by 11 scholars, 129–205.

Gilbert, J.P. and Mosteller, F. (1972). The Urgent Need for Experimentation. *On Equality of Educational Opportunity*, edited by F. Mosteller and D. Moynihan. Random House, New York.

Hall, J.A., and Taylor, M.C. (1985). "Psychological Androgyny and the Masculinity X Femininity Interaction," *J. of Personality and Social Psychology*, Vol. 49, 429–435.

Harrington, D.M., and Anderson, S.M. (1981). "Creativity, Masculinity, Femininity, and Three Models of Psychological Androgyny," *J. of Personality and Social Psychology*, Vol. 41, 744–757.

Heid, M.K. (1988). Resequencing Skills and Concepts in Applied Calculus Using the Computer as a Tool. *J. of Research in Math Ed.*, Vol. 19, No. 1, 3–25.

Heilbrum, A.B., and Schwartz, H.L. (1982). "Sex-Gender Differences in Level of androgyny," *Sex Roles*, Vol. 8, 201–213.

Hicks, C.R. (1982). *Fundamental Concepts in the Design of Experiments*, 3rd ed. Holt, Rinehart, and Winston, New York.

Kalin, R. (1979). "Method for Scoring Androgyny as a Continuous Variable," *Psychological Reports*, 44, 1205–1206.

Kruskal, W. (1988). Miracles and Statistics: The Causal Assumption of Independence. *JASA*, Vol. 83, No. 404, 929–940.

Lipta, R. (1985). "Review of Bem Sex-Role Inventory," *Psychological Review*, 176–178.

Lubinski, D., Tellegen, A., and Butcher, J.N. (1981). "Relationship between Androgyny and Subjective Indicators of Emotional Well-being," *J. of Personality and Social Psychology*, Vol. 40, 722–730.

Lubinski, D., Tellegen, A., and Butcher, J.N. (1983). "Masculinity, Femininity, and Androgyny Viewed and Assessed as Distinct Concepts," *J. of Personality and Social Psychology*, Vol. 44, 428–439.

Marquardt, D.W. (1963). "An Algorithm for Least-Squares Estimation of Nonlinear Parameters," *J. for the Society of Industrial and Applied Mathematics*, Vol. 11, 431–441.

Mosteller, F. (1981). Innovation and Evaluation. *Science*, Vol. 211, 881–886.

Nelder, J.A. (1986). "Comment to Hasti and Tibshirani's paper 'Generalized Additive Models,'" *Statistical Science*, Vol. 1, 312.

Patton, M.Q. (1980). *Qualitative Evaluation Methods*. Sage, Beverly Hills.

Shostrom, E.L.(1980). *Personal Orientation Inventory*. Educational and Industrial Testing Service, San Diego.

Spence, J.T., Helmreich, R.L., and Stapp, J. (1975). "Ratings of Self and Peers on Sex-Role Attributes and their Relation to Self-Esteem and Conceptions of Masculinity and Femininity," *J. of Personality and Social Psychology*, Vol. 32, 29–39.

Chapter 6

On Objectivity, Subjectivity, and Probability

In this chapter we will devote ourselves to certain aspects of objective science and subjective knowledge. For this purpose, we begin with an appreciation of the advantages and limitations of logic and mathematics—the unchallengeable "objective science."

The first advantage of logic is that reasoning alone is capable of leading to new discoveries—despite a common belief that deductive reasoning produces no new knowledge. Here is an example of how scientists are able to reach for new discoveries by manipulating abstract symbols.

According to Einstein's special theory of relativity, the energy of a particle that has mass m and momentum p is given by

$$E^2 = m^2c^4 + p^2c^2. \tag{1}$$

If the momentum p is zero, the equation reduces to the well-known

$$E = mc^2.$$

By taking the square root of the equation (1), mathematically we should have

$$E = mc^2 \quad \text{and} \quad E = -mc^2.$$

Unfortunately, negative energy was considered as not supported by physical reality and was dismissed by most physicists, Einstein included (Gribbin, 1984).

In the late 1920s, Paul Dirac formulated a new wave equation for electrons, which incorporated relativity into quantum mechanics. Dirac's equation is very complicated, but it explains the spin of the electron as an angular momentum arising from a circulating flow of energy and momentum within the electron wave. His equation also predicts a relationship between the magnetic moment and the spin of the electron (Ohanian, 1987). However, Dirac's monumental work was considered by most physicists as marred by one "defect": he attempted to explain the negative solution in his equation by the existence of particles in a sea of negative-energy state.

Dirac's theory of negative-energy particles was born out of a conviction that his complicated calculations were perfect in almost every aspect and thus should not result in producing a nuisance negative number. Therefore he invented negative-energy particles to account for an "imperfect feature" in his equation. But his theory was completely anti-intuitive and no physicist could take the idea seriously. In 1932, as a surprise to everybody, Carl Anderson discovered the tracks of electronlike particles of positive charge in a cloud chamber. Subsequent measurements concluded that the new particles have the same mass, same spin, and same magnetic moment as electrons—they are identical with electrons in every respect, except for their electric charge. The new particle was later called the anti-electron. In the mid-1950s, the discovery of anti-proton and anti-neutron further confirmed Dirac's original ideas that were mainly based on the faith of mathematical perfection. (Anti-matters are now routinely used in particle accelerators that gobble up billions of dollars of federal money and involve hundreds of physicists.)

Dirac was known as a man whose only scholarly activity was playing with mathematical equations. In 1933 he was honored with a Nobel Prize for his equation and a bold idea that literally opened up a new universe for human-kind. His story also indicates that fooling around with mathematical equations may pay off in totally unexpected ways.

I. STATISTICAL "JUSTIFICATION" OF SCIENTIFIC KNOWLEDGE AND SCIENTIFIC PHILOSOPHY

The second advantage of deductive reasoning is that if a theorem is proved true, then the conclusion of the theorem will never change—it is an ultimate truth. If one desires pure objective knowledge, then one has to re-examine mathematics for its ability to produce unbending knowledge. In the early part of the 20th century, logical positivism did just that. The movement also strived to develop *the logic of science* (Carnap, 1934) and to "propagate a scientific world view" (Schlick, 1928). A major goal of the movement was to establish a "scientific philosophy," in contrast to the traditional "speculative philosophy" of Plato, Socrates, Descartes, Hume, etc. (Reichenbach, 1951).

The movement soon dominated the field of the philosophy of science. One of their tasks was to reduce all objective knowledge to formal logic. A monumental achievement in this direction was the publication of *Principia Mathematica*, Volumes I–III (Russell and Whitehead, 1910–1913). Russell's goal was to develop a symbolic logic and to show that all pure mathematics can be reduced to formal logic. The attempt was driven by a desire to eliminate contradictions that threaten the foundations of set theory and other branches of mathematics. It was believed that mathematics and logic are inseparable, and that if we want to obtain purely objective knowledge, then first we have to reconstruct mathematics as a body of knowledge that is contradiction-free (Davis and Hersh, 1981).

Fig. 1 is an example of Russell's legendary work on how the arithmetic proposition $1 + 1 = 2$ is established (after 362 pages of elaboration). This mammoth exercise is likely to be viewed by non-mathematicians as plain stupidity. But it is an attempt (a praiseworthy attempt) to establish mathematics as a

$$*54\cdot42. \quad \vdash::\alpha\epsilon2.\supset:.\beta\subset\alpha. \quad !\beta.\beta\neq\alpha. \equiv.\beta\epsilon\iota``\alpha$$

Dem.

$$\vdash.*54\cdot4. \quad \supset\vdash::\alpha=\iota'x\cup\iota'y.\supset:.$$

$$\beta\subset\alpha.\exists!\beta. \equiv.\beta=\Lambda.v.\beta=\iota'x.v.\beta=\iota'y.v.\beta=\alpha:\exists!\beta:$$

$$[*24\cdot53\cdot56.*51\cdot161] \qquad \equiv:\beta=\iota'x.v.\beta=\iota'y.v.\beta=\alpha \qquad (1)$$

$$\vdash.*54\cdot25.\text{Transp}.*52\cdot22.\supset\vdash:x\neq y.\supset.\iota'x\cup\iota'y\neq\iota'x.\iota'x\cup\iota'y\neq\iota'y:$$

$$[*13\cdot12] \quad \supset\vdash:\alpha=\iota'x\cup\iota'y.x\neq y.\supset.\alpha\neq\iota'x.\alpha\neq\iota'y \qquad (2)$$

$$\vdash.(1).(2). \quad \supset\vdash::\alpha=\iota'x\cup\iota'y.x\neq y.\supset:.$$

$$\beta\subset\alpha.\exists!\beta.\beta\neq\alpha. \equiv:\beta=\iota'x.v.\beta=\iota'y:$$

$$[*51\cdot235] \qquad\qquad \equiv:(\exists z)\cdot z\epsilon\alpha.\beta=\iota'z:$$

$$[*37\cdot6] \qquad\qquad \equiv:\beta\epsilon\iota``\alpha \qquad (3)$$

$$\vdash.(3).*11\cdot11\cdot35.*54\cdot101.\supset\vdash.\text{Prop}$$

$$*54\cdot43. \quad \vdash:.\alpha,\beta\epsilon1.\supset:\alpha\cap\beta=\Lambda. \equiv.\alpha\cup\beta\epsilon2$$

Dem.

$$\vdash.*54\cdot26.\supset\vdash:.\alpha=\iota'x.\beta=\iota'y.\supset:\alpha\cup\beta\epsilon2. \equiv.x\neq y.$$

$$[*51\cdot231] \qquad\qquad\qquad \equiv.\iota'x\cap\iota'y=\Lambda.$$

$$[*13\cdot12] \qquad\qquad\qquad \equiv.\alpha\cap\beta=\Lambda \qquad (1)$$

$$\vdash.(1).*11\cdot11\cdot35.\supset$$

$$\vdash:.(\exists x,y).\alpha=\iota'x.\beta=\iota'y.\supset:\alpha\cup\beta\epsilon2. \equiv.\alpha\cap\beta=\Lambda \qquad (2)$$

$$\vdash.(2).*11\cdot54.*52\cdot1.\supset\vdash.\text{Prop}$$

From this proposition it will follow, when arithmetical addition has been defined, that $1 + 1 = 2$.

model of absolute truth. The attempt was indeed a grand achievement of axiomatic reasoning. It also inspired the great mathematician David Hilbert to put forth a program that he claimed would banish all contradictory results in mathematics. In retrospect, Hilbert's faith in this program was admirable: "What we have experienced twice, first with the paradoxes of the infinitesimal calculus and then with the paradoxes of set theory, cannot happen a third time and will never happen again" (Regis, 1987).

In 1931, six years after Hilbert's manifesto, Kurt Godel proved, by the very method of logic, that all consistent axiomatic formulations of number theory include undecidable propositions. In plain English, the theorem showed that no logical system, no matter how complicated, could account for the *complexity* of the natural numbers: 0, 1, 2, 3, 4, ... (Hofstadter, 1979). The theorem thus puts an end to the search for logical foundations of mathematics initiated by Russell and Hilbert.

After decades of elaboration on the logic of science and on "the ultimate meaning of scientific theories" (Schlick, 1931, p. 116), Russell (1948) finally conceded that "all human knowledge is *uncertain*, inexact and partial."[1] This conclusion may be true, but the tone is too pessimistic. It may also lead to the anarchism that nothing is knowable and therefore anything goes. For instance, when a space shuttle blows up in the sky and kills all of the astronauts in it, the scientists in charge cannot simply shrug off the event by saying that "all human knowledge is uncertain, inexact and partial."

Moving back to mathematics and logic, although Godel had proved that logical systems are intrinsically incomplete, he never proved that mathematics as a branch of human knowledge is inconsistent or uncertain.

It should be spelled out that, outside the domain of pure mathematics, there are two types of knowledge: local and global knowledge. While local knowledge can be quite certain, global knowledge is in general partial and hard to justify. For example, when Newton released an apple from his hand, the apple dropped to the floor. That is certain. But what is the reason free objects all drop to the ground? According to Aristotle, that is the fate of everything. According to Newton, that is the result of gravity. According to Einstein, that is the result of relativity.

Given the competing theories, a big question is, how do we know which one is the best? Furthermore, a theory is a generalization of finitely many observations. What is the scientific justification of this generalization? According to logical positivists, the answer is the assignment of a *degree of confirmation* to each scientific theory. This so-called "degree of confirmation," according to Reichenbach (and his Berlin school of philosophy), rests upon the calculus of probability, which is the "very nerve of scientific method." Also, the notorious Hume's problem of induction "can thus be solved" (Reichenbach, 1951).

In addition, when conflicting theories are presented, probability and "inductive logic" are used to settle the issue. More precisely, "the inductive inference is used to confer upon each of those theories a degree of probability, and the most probable theory is then accepted" (Reichenbach, 1951). This constitutes the core of the "scientific philosophy" advocated by Reichenbach and his Berlin school of philosophy.

However, the justification of scientific knowledge is often not an easy matter, let alone the justification of "scientific philosophy." For example, what is the probability that a specific nuclear-power plant will fail? In such a tightly controlled situation, if scientists cannot come up with a reliable probability, then there is little hope of assigning a meaningful probability to each scientific theory. As a matter of fact, many famous statisticians have plunged into the problem of the nuclear-power plant only to retreat with disgrace. (See Breiman, 1985; Speed, 1985).

It is now clear that the "speculative philosophy" of Plato, Descartes, and Hume contributed more to human knowledge than Reichenbach's voluminous writings on the so-called "scientific philosophy."

Nevertheless, the influence of logical positivism has been enormous in many branches of science in such distinctive manners as: (1) using mathematical probability to justify certain unjustifiable "scientific" results; (2) masquerading "subjective knowledge" under the cover of "objective estimation."

This kind of unfortunate influence poses less problem in the natural sciences than in social-behavioral studies. One reason for this phenomenon is that the natural sciences are usually built upon a solid foundation and investigators are too busy to worry about philosophical debate on the nature of their research conducts.

Social and behavioral scientists, on the other hand, are insecure about their research findings. Therefore, they need "justification" like the one advocated by Reichenbach. This justification, as we have seen in the previous chapters, is misleading and is detrimental to their research efforts.

●   ●   ●   ●   ●   ●   ●   ●   ●

In general, it is easier to develop the scientific theories of nature than to grasp the nature of scientific theories. But fortunately, Thomas Kuhn (1970), a physicist, provided us with a better picture in this regard. According to Kuhn, scientists work under different paradigms that are "universally recognized scientific achievements that for a time provide model problems and solutions to a community of practitioners." Under a paradigm, a scientific community devotes itself to solving problems that are narrowly focused or fragmented, or even problems that apparently need not be solved.

If you reflect upon it, isn't this exactly what we have been trained to do and have been happily doing throughout our professional lives?

Kuhn characterized the above state of problem-solving as the period of "normal science." During this period, scientific investigations, albeit limited, can be highly productive. Nevertheless, he maintained that the greatest discoveries are not made during the period of normal science, but rather during periods of paradigmatic revolutions.

Paradigmatic revolutions occur, according to Kuhn, when "anomalies" accumulate. That is, when practitioners of a developed science find themselves unable to solve a growing number of problems. This leads to crisis and in turn provides the opportunity for the rise of a new paradigm.

Kuhn's insight was truly revolutionary and soon became a new paradigm itself in the philosophy of science. There is no denying that much of what Kuhn had described about scientists and scientific activities is true. It is also to his credit that he put human beings back into a global picture of the scientific enterprise. This is what it is and this is what it should be.

However, Kuhn went one step further to question the very methods of scientific investigation. For instance, he wrote,

> [During my year-long visit at the Center for Advanced Studies in the Behavioral Sciences], I was struck by the number and extent of the overt disagreements between social scientists about the nature of legitimate scientific problems and *methods*. Both history and acquaintance made me doubt that practitioners of the natural sciences possess firmer or more permanent answers to such questions than their colleagues in social science. [emphasis supplied]

Furthermore, Kuhn maintained that there is no such thing as progress in science. For instance, Kuhn (1970, pp. 206–207) wrote:

> I do not doubt, for example, that Newton's mechanics improves on Aristotle's and that Einstein's improves on Newton's as instruments for puzzle-solving. But I can see in their succession no coherent direction of ontological development. On the contrary, in some important respects, though by no means in all, Einstein's general theory of relativity is closer to Aristotle's than either of them is to Newton's.

Such views on science run against the theme of this book. Therefore, we have no other choice but to put forward our assessment of his conclusions. In the case of Einstein's theory, we believe that he was using a half truth to advocate a radical view that is often used by social scientists in the defense of their sloppy statistical work. (See Freedman, 1987.)

The truth about "Einstein's theory" is that there are actually two theories: the general theory and the special theory of relativity; one deals with gravity while the other does not. Kuhn's assertion is that, from a paradigmatic point of view, the general theory is "closer to Aristotle's than either of them is to Newton's." The reasons, to my understanding, are as follows.

A distinctive feature of the Einstein paradigm is that many of the new theories based upon general relativity are just "theories": they are not empirically testable, and the way theorists are pumping up suppositions is very much in the manner Aristotle framed up his views of nature.

For instance, in order to test the super-string theory or certain Grand Unified Theories, the particle accelerators would have to be light years across in size. This is hardly in the spirit of Newtonian physics, whose trademark was firmly established when Galileo started doing experiments to test Aristotle's claims on falling objects. But now with these new theories about Grand Unification, physicists are just sitting there thinking about exotic theories and not doing experiments.

Another prominent feature of general relativity is that it is not a well-tested theory itself, at least not until recent years. As a matter of fact, throughout his life, Einstein saw only two confirmations of the general theory of relativity. Both confirmations rely on observational data, not controlled experiments. In addition, the confirmations are distinctively *statistical*. Together, the confirmations constitute two case studies about the troublesome nature of statistical analysis that are common in non-experimental sciences.

One confirmation of general relativity is about the peculiar movement of Mercury, a strange phenomenon that cannot be explained by Newton's law of gravity. Numerous proposals were put forward to account for the discrepancy. But it was Einstein's framework that appeared to explain the movements of Mercury and other planets in the most elegant fashion. Table 1 shows some statistics in this respect (Marion, 1976, p. 294):

Table 1   Precessional Rate (in arcseconds per century)

Planet	Theoretical calculation	Observed
Mercury	43.03 ± 0.03	43.11 ± 0.45
Venus	8.63	8.4 ± 4.8
Earth	3.84	5.0 ± 1.2

But just like any statistical artifact in observational studies, the confirmation was challenged by other observations. In 1966 Dicke and Goldenberg found that the Sun is not a sphere and that the flatness of the Sun contributes about 3 arcseconds to Mercury's strange movement. This observation cast doubt on general relativity and seemed to favor other competing theories such as the Brans-Dicke theory. (For more details, see, e.g., Will, 1986.) But then others soon pointed out that the Dicke-Goldenberg results might be due to the instrinsic solar brightness between poles and equator, instead of a true flatness of the Sun. Debates surrounding these matters abounded in the scientific literature. Today, scientists appear confident that the issue will eventually be resolved in favor of general relativity. But so far no decisive piece of evidence has been produced.

Another confirmation of general relativity was the well-touted Eddington expedition of 1919. The event has been argued by some (e.g., the British historian Paul Johnson)[2] as the beginning of the "modern era," in that the old concepts of absolute space and absolute time were lost forever.

But the fact is that Eddington's confirmation is somewhat scandalous. Here we will relate the story as appeared in Earman and Glymour (1980) and in Fienberg (1985). We will also give some of our calculations and comments.

According to the general theory of relativity, starlight is bent by the gravitational field of the Sun. Einstein's theory predicted a deflection of 1.74″, as opposed to the value of 0.87″ predicted by Newtonian theory.[3] To settle the issue, Eddington organized a pair of expeditions to make appropriate measurements during the eclipse of the sun. Three sets of data were collected (all measurements are in arcseconds):

Data set	Mean deflection	Probable error
1	1.61″	0.444″
2	0.86″	0.48″
3	1.98″	0.178″

As anyone can tell, the second data set yielded a mean deflection that is almost identical to the Newtonian value of 0.87″. But this set of data was *not* reported when Eddington returned to an extraordinary joint meeting of the Astronomical and Royal Societies. Einstein's prediction was thus "confirmed" and a modern era of curved space and time was born.

In a critique of the Eddington expedition, Earman and Glymour (1980) pointed out many troublesome features in the whole affair. To begin with, it is very tricky (or error-prone) to measure what ought to be measured during an eclipse. For example, comparison photographs were taken at different times, one set during the day and the other at night. Temperature changes, the relocation of the telescope, and slightest mechanical deviations in the machine could have, in theory and in practice, significant impact on the measurements.

For another example, the first data set in the table were calculated from the images of only two plates. Further, the images were so blurred that Eddington disregarded the standard least-squares technique and expended extra effort to derive a scale factor indirectly from other check plates. In addition, the rotational displacement of the images was obtained by assuming the validity of Einstein's predicted deflection.

Behind the stage, the way Eddington deleted or adjusted his data does not appear to be a model of scientific conduct. Nevertheless, the Eddington "confirmation" of Einstein's theory is, by any standard, a great success in media and in the scientific community. The "confirmation" indeed made the name Einstein a household word.[4]

After the 1919 confirmation, there were other expeditions to test Einstein's theory. The results varied, but most of them were a bit higher than the deflection predicted by the general relativity. In their comments about these confirmations, Earman and Glymour (1980) wrote, "It mattered very little. The reputation of the general theory of relativity, established by the British eclipse expeditions, was not to be undone." To support this position, Earman and Glymour related the affair to the prediction of gravitational red shifts.

The red shift was the first of famous predictions based on the notion of curved space-time. The prediction was made in 1907 by Einstein himself. The confirmations of this prediction, however, had always been elusive. Before 1919, no one claimed to have obtained red shifts of the size predicted by the theory. But within a year of the announcement of the Eddington results several researchers reported finding the Einstein effect. According to Earman and Glymour (1980),

> there had always been a few spectral lines that could be regarded as shifted as much as Einstein required; all that was necessary to establish the red shift prediction was a willingness to throw out most of the evidence and the ingenuity to contrive arguments that would justify doing so.

In addition, Earman and Glymour maintained:

> The red shift was confirmed because reputable people agreed to throw out a good part of the observations. They did so in part because they believed the theory; and they believed the theory, again at least in part, because they believed that the British eclipse expeditions had confirmed it.

Such criticisms, amazingly, have not eclipsed the fame or the "validity" of general relativity. Modern observations, according to certain experts in the field (see, e.g., Hawking, 1988, p. 32), have accurately confirmed certain predictions; and the theory so far still holds the truth about space, time, and gravity. But other physicists, such as Kuhn, may not be as convinced as these experts.

We now turn to a comparison between general relativity and special relativity. First, special relativity deals specifically with physics in the situations where gravitational force is negligible. In addition, unlike general relativity, special relativity has been checked and rechecked and confirmed time and time again (by controlled experiments). Indeed the theory has been accepted by physicists as a reality beyond any shadow of doubt. For instance, just by the sheer power of the atomic bomb, you don't want to argue with the equation $E = mc^2$ (Will, 1986).

Second, special relativity gave rise to Dirac's theories about the spins of elementary particles and about the existence of anti-matters, all of which have been confirmed in tightly controlled experiments over and over again.

Third, special relativity also paved the way for the theory of quantum electro-dynamics (QED), the most accurate theory in all of physics (Ohanian 1987, p. 444; Will, 1986).

Kuhn himself is an expert in physical science. It is thus a big surprise that he did not credit the coherent progress in physical science exemplified in the ontological development from Newtonian physics and special relativity to Dirac's theory and the theory of quantum electro-dynamics.

Nevertheless, his brilliant observation about general relativity reinforced a suspicion that in certain branches of non-experimental (and semi-experimental) science, "progress," despite mountainous publications, might be less than it seems.

II. CLASSICAL PROBABILITY, COMMON SENSE, AND A STRANGE VIEW OF NATURE

In the previous section we discussed some advantages of deductive reasoning. In this section we will further discuss another advantage, and later in this section the limitations of logic and mathematics. The advantage is that mathematical reasoning often produces results that are contradictory to our intuition and common sense. The clash of mathematics and intuition, to my knowledge, always results in the surrender of intuition to mathematical reasoning—because experiments always turn out on the side of logical reasoning. To illustrate this point, consider the following example:

> There are three fair dice whose 3×6 sides will be labelled from 1 to 18 by Mr. A (a mathematician). After the dice are labelled, Mr. B (another mathematician) chooses one of the dice. Finally Mr. A chooses another die from the remaining ones. The two players roll their dice and the person with the larger number wins $200.

In this game, our intuition says that in the long run Mr. B will have advantage over Mr. A, because Mr. B has the first choice of die. Careful calculation says that our intuition is wrong. For instance, if Mr. A arranges the 18 numbers as follows: I = {18,10,9,8,7,5}, II = {17,16,15,4,3,2}, and III = {14,13,12,11,6,1}. Then

$$Pr[I > II] = (6 + 3 + 3 + 3 + 3 + 3)/36 = 21/36,$$

$$Pr[II > III] = (6 + 6 + 6 + 1 + 1 + 1)/36 = 21/36,$$

$$Pr[III > I] = (5 + 5 + 5 + 5 + 5 + 1)/36 = 21/36 > 50\%.$$

In other words, die I is better than die II; die II is better than die III; and die III is better than die I. Hence Mr. A will always beat Mr. B, no matter how smart Mr. B is![5] This may sound impossible. If you don't believe this conclusion, we can go somewhere and bet.

The previous dice are capable of producing more surprises. For example,

$E(I)$ = the mathematical expectation of rolling die I

$= (18 + 10 + 9 + 8 + 7 + 5)/6 = 9.5 = E(II) = E(III)$.

That is, the expected values of the three dice are the same, even if we have just concluded that the probability die I will beat II is larger than 50%.

If you have enjoyed the above example, the next will be equally fascinating:

Game 1: Flip a coin; the gambler will bet $1 on either head or tail (the probability of winning is .5).

Game 2: the gambler will bet $1 either red or black on a roulette (the probability of winning is 18/38, or .474).

Assume that in game 1 the gambler has $900 and his goal is $1,000,000 (one million), and that in game 2 the gambler also has $900 but his goal is only $1,000 (one thousand). In which game is the gambler more likely to reach his goal? Game 1 or 2?

In graphic form, the above information can be summarized as follows:[6]

Game 1: $900 $\longrightarrow$ $1,000,000   (p = .50)

Game 2: $900 $\longrightarrow$ $1,000   (p = .474 $\cong$ .50)

Intuitively, it is a long way to go from $900 to one million dollars. Therefore it is more likely to reach the goal in Game 2 than Game 1. However, if we let

$h(K)$ = Pr[reaching the goal of $N, giving that the initial capital = $K],

then standard calculations of Markov absorption probabilities yield the following formulas:

Game 1: $h(K) = K/N$, N = $1,000,000   (1)

Game 2: $h(K) = \dfrac{(q/p)^{\wedge}K - 1}{(q/p)^{\wedge}N - 1}$, $q = 1 - p$,   N = $1000   (2)

If K = $900, we obtain the following probabilities:

Game 1: $h(900)$ = .09%

Game 2: $h(900)$ = .003%

Note that $.09/.003 = 30$. This means that the gambler is 30 times more likely to reach the goal in Game 1 than in Game 2, a conclusion quite contrary to intuition for most people.

In their book, *How to Gamble If You Must?*, Dubins and Savage (1965) mentioned a strategy that can be applied to the situation of Game 2: Bet $100 the first time; if you lose, then bet $200 the second time, etc. Using this strategy,

$$h(900) = p + \frac{pq + p(q\hat{\ }2)}{1 - (pq)\hat{\ }2} = 88\%$$

In other words, if one uses the Dubins-Savage strategy in Game 2, then the probability he will reach the goal is about 88%. Without any strategy, the probability is .00003.

It now appears that the Dubins-Savage strategy is a sure way to win in the long run. But this is not true. For instance, if we calculate the expected values, the answer is −$20 with the strategy and −$899.97 without a strategy. These negative values lead to a question: In the long run, which is better? to play with or without using the Dubins-Savage strategy? To answer this question, we let

> m(K) = the expected time the gambler either reaches his goal or gets ruined.

Then for Game 2 (without strategy),

$$m(K) = 1 + p * m(K + 1) + q * m(K - 1); \quad K = 1, 2, \ldots, N - 1;$$
$$m(0) = 0, m(N) = 0.$$

The solution to these equations yields

$$m(K) = 22492 \approx 22{,}500.$$

This means that it will take about 22,500 spins of the roulette wheel for the gambler (without strategy) to reach his goal or get ruined. On the other hand, it takes, on the average, only 2 spins to end a game if one uses the Dubins-Savage strategy. Recall that the expected gain of the Dubins-Savage strategy is −$20. Therefore,

$$(-\$20) * \frac{22500 \text{ spins}}{2 \text{ spins}} = -\$225{,}000$$

In other words, the gambler who uses the Dubins-Savage strategy repeatedly will lose about $225,000 by the time the other gambler (without strategy) loses $900.[7]

Similar puzzles are plentiful in probability and many branches of mathematics (see, e.g., Coyle and Wang, 1993). These puzzles may appear "contradictory" to novice students or amateur scientists. If well explained, these seemingly contradictory results are consistent; more importantly, they are supported by experiment. In other words, the "paradox" is only a conflict between reality and your feeling of what reality ought to be. This is one of the most valuable assets of mathematics over common-sense reasoning.

We now turn to the limitation of logic and mathematics. In human reasoning, the major problem with logic is that logic itself cannot check the assumptions in a chain of arguments. To illustrate, consider the following example:

A pencil has five sides, with only one side carved with a logo. If you toss the pencil, what is the chance that the logo will come out on top?

So far almost all answers I have heard are the same: 1/5. The correct answer is 0 or 2/5, depending on the definition of "top." In this example, no amount of logic or mathematical formulas can help to reach the correct answer unless the investigator looks at the pencil in his hand and thinks about the pencil in the stated problem.

The moral of this example is that a total reliance on logic is not enough. This may sound like an old cliché, but isn't much of the activity of logical positivism merely using logic to justify all human knowledge? The following example is personal, but it is quite typical. A group of Ph.D. candidates from Asia were involved in a debate on the political future of their country. After heated exchange, a mathematics candidate who was trained in doing abstract algebra lectured another anthropology candidate: "I am a mathematician. I am trained in logic. So my conclusion is more scientific than your conclusion."

In my observation, many *quantitative* researchers are like that naive mathematics candidate—they believe that pure logic (e.g., statistical formulas and electronic computers) will produce more scientific results than anectodal evidence. Worse, they are very rigid about the "rigor" and the "objectivity" of their statistical results. Since they have learned the wrong things too well, we will put forth additional criticism of such mentality in Sections III, IV, and V of this chapter.

•   •   •   •   •   •   •   •   •

In the previous example of tossing a pencil, the question can be more complicated:

Toss the pencil 100 times. What is the probability that the logo will show up at least 30 times?

In this case, one can first examine the shape of the pencil and then apply standard mathematics to solve the problem. However, in many scientific applications of the probability calculus, the phenomenon under study is not as easy as the shape of a pencil. For example, the assumptions in a social-behavioral regression often do not hold water, but researchers seldom hesitate to calculate the P-values of the coefficients and declare that the equation is "scientific" and "objective."

Well, if you blindfold yourself, the statistics generated by a fixed procedure are certainly "objective." But "objectivity" does not by itself guarantee "reliability," as the five-sided pencil has taught us.

In the next example, we will examine a case where the shape of the thing is so uncertain that the very foundation of classical probability theory is challenged. In addition, this case study is intended to illustrate the fact that mathematical statistics is such a rich field that it is capable of producing more surprising results.

According to the standard theory, the conditional probability of event A to happen, given that event B already happened is

$$Pr[A|B] = \frac{Pr[A \text{ and } B]}{Pr[B]} \tag{3}$$

and the total probability of A can be split into two distinct parts:

$$Pr[A] = Pr[A \text{ and } B1] + Pr[A \text{ and } B2], \tag{4}$$

provided that $\{B_1, B_2\}$ is a partition of the sample space. By the ground rules of formulas (3) and (4), it is straight-forward to derive the following:

$$Pr[A|C] = Pr[A|B1 \text{ and } C] * Pr[B1|C] + Pr[A|B2 \text{ and } C] * Pr[B2|C]. \tag{5}$$

This Bayes formula looks very simple, but its implication is mind-blowing. Consider the experiment shown in Fig. 2. In this set-up, the left-hand side of formula (5) does not equal the right-hand side, because the former reveals the wave property, while the latter corresponds to the particle property of the electron. On the left-hand side, $Pr[A|C]$, we have no knowledge about whether the electron passes through slit 1 or 2. On the right-hand side, $Pr[A|Bi \text{ and } C]$ indicates that we know the electron passes through slit i.

The two-slit experiments are the most famous examples in modern physics,[8] although most physics books do not discuss the Bayes formula in (5). Here are some implications provided by Richard Feynman (1948): "(5) results from the

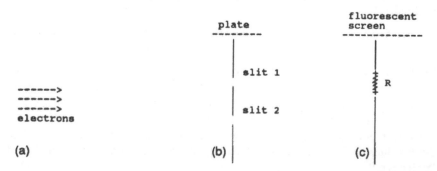

Figure 2 (a) The event that an electron falls in region R; (b) the event that the electron passes through slit i, i = 1 or 2; (c) the event corresponding to the energy level of the electron.

statement that in each experiment, B had some value. The only way (5) could be wrong is that the statement 'B had some value' must sometimes be meaningless." In the *Berkeley Symposium on Mathematical Statistics and Probability*, Feynman (1951) further concluded: "It is not true that the electron passes *either* through hole 1 *or* hole 2." But in reality, electrons always have the same mass, same charge, same spin, and same magnetic moment. One-half electrons do not exist.

This parodox is the kernel of the orthodox view on elementary particles. A majority of physicists such as Feynman, Heisenberg, and Bohr all share this view. Amused by this orthodox interpretation, a physicist in the Far East put a sign on his office door: "Richard Feynman might have been here."

In 1964, John Bell (a European physicist) discovered an inequality that greatly strengthens the orthodox view. The inequality and its experimental verifications are too lengthy to be reviewed here. (See, e.g., Shimony, 1988.) But a theorem derived from the inequality is easy to state:

THEOREM   (J. Bell, 1965)

> There exist vectors a, b, c in the 3-dimensional Euclidean space for which it is *not possible* to find 3 random variables S(a), S(b), S(c) on a probability space with values in $\{+1, -1\}$ such that their correlations are given by
>
> $$E[S(x)S(y)] = xy,$$
>
> for all x, y = a, b, c.

Bell's inequality and related theorems lend strong support to the following orthodox views where truth seems stranger than fiction (Arcardi, 1985):

> An electron is a wave packet which, when nobody looks at it, is virtually everywhere in the space. (This is called the *superposition state* of electron.)
>
> However, as soon as somebody looks at it, the electron suddenly materializes in a region of a few microns. (This is part of a theory called *collapsing waves*.)
>
> No probability distribution can describe the wave packet before you look at it. (This is called the *non-locality* of one electron.)
>
> Before you look at them, two electrons interact with each other. There is no way to tell them apart. (This is called the *non-separability* of two electrons.)

In 1935, Einstein and co-workers published a famous article against the popular interpretation of electron. His main argument is that the non-separability of two electrons implies an instantaneous "communication" between two electrons at a distance. This communication is faster than the speed of light and thus violates a fundamental principle firmly established in physics.

Einstein's article was intended to point out an impossibility, but it turned out to be prophetic. Experiments (see, e.g., Shimony, 1988, *Scientific American*) appear to have strong evidence that two entities separated by many meters can exhibit striking correlations in their behavior, so that a measurement done on one of the entities seems instantaneously to affect the result of a measurement on the other.

On another front, Schrödinger challenged the theory of "collapsing waves" by a thought experiment now widely known as "Schrödinger's cat." The example neatly penetrates the weak spot of the orthodox view. Imagine that a cat is sealed in a box, along with a Geiger counter, a small amount of radioactive material, a hammer, and a bottle of poison (such as cyanide). The device is arranged so that when an atom of the radioactive substance decays, the Geiger counter discharges, causing the hammer to crush the bottle and release the poison (Gribbin, 1984).

According to the orthodox interpretation, before we look inside the box, nothing can be said about the radioactive sample—it is not true that the sample *either* decayed *or* did not decay, because the wave function collapses only when we look at it. But if the state of the radioactive sample is indeterminate (before we look at it), so are the bottle of poison and the cat. In other words, the radioactive sample has both decayed and not decayed, the bottle is both broken and not broken, and the cat is both dead and alive.

This thought experiment (as well as other parodoxes in quantum mechanics) has generated numerous scholarly acticles and Ph.D. dissertations. But the whole issue is far from settled. Among the different proposals, the most interesting ones are put forth by a group of Italian theorists who challenge directly the validity of using standard probability theory in the micro-universe. Specifically, they challenge the validity of the conditional probability in formula (3) and thus the Bayes formula in (5). In doing so, they propose non-Kolmogorovian probability model for micro phenomenon. The ground of their proposal is that the very act of "looking at it" means an interference of light (photons) with electrons. Therefore the Kolmogorovian model of probability may not hold in the micro-universe. Here is one of the theorems in this new direction.

THEOREM   (L. Accardi, 1982)

Let p, q, r be real numbers in the open interval $(0,1)$.
Let

$$P = \begin{bmatrix} p & 1-p \\ 1-p & q \end{bmatrix}, \quad Q = \begin{bmatrix} q & 1-q \\ 1-q & q \end{bmatrix}, \quad R = \begin{bmatrix} r & 1-r \\ 1-r & r \end{bmatrix}.$$

Then

1. A Kolmogorovian model for P, Q, R exists if and only if

$$|p + q - 1| < r < 1 - |p - q|.$$

2. A complex Hilbert space model for P, Q, R exists if and only if

$$1 < \frac{p + q + r - 1}{2(pqr)^{1/2}} < 1.$$

3. A real Hilbert space model for P, Q, R exists if and only if

$$\frac{p + q + r - 1}{2(pqr)^{1/2}} = \pm 1.$$

4. A quaternion Hilbert space model for P, Q, R exists if and only if a complex Hilbert space model exists.

Accardi and his co-workers believe that Kolmogorovian model is not universally applicable, and that parodoxes created by orthodox views of quantum mechanics may eventually be eliminated by Hilbert space models. Their works are very interesting, but their conviction so far has not been widely accepted.

Historically, physics was part of philosophy. The two separated when Galileo used experimental methods to test doubtful claims initiated by Aristotle. By this standard, the rising non-Kolmogorovian models of quantum mechanics have to be deemed philosophical. If the models are further developed and tested by a rich array of experimental data, then the models eventually may become part of physics.

To the insiders of academic endeavors, science is not always logical. It is interesting to see how physicists live with all the uncertainties and parodoxes surrounding the foundation of their discipline. Feynman, for instance, feels perfectly comfortable with the uncertainty. He deserves it. His discipline is so far the most solid and accurate among all natural sciences. Further, his QED (quantum electrodynamics) stands as the most accurate theory in all physics. In short, if your discipline has good theory, good measurement, and accurate prediction, then you are entitled to be a little arrogant, like Richard Feynman.

III. INTUITION AND SUBJECTIVE KNOWLEDGE IN ACTION: THE BAYES THEOREM (AND ITS MISUSE)

One of the tasks of logical positivism is to outlaw all speculative statements as meaningless. A consequence of this "scientific philosophy" is a disrespect of intuitive and subjective knowledge. In statistics, this influence is reflected by the strict teaching of the Neyman-Pearson school of statistical reasoning. In social-behavioral sciences, this influence also "led to the declaration that concepts like mind, intuition, instinct, and meaning were 'metaphysical' survivals unworthy of study" (Harris, 1980, p. 15).

Under the shadow of both logical positivism and Neyman-Pearson's rigid teaching, certain researchers are afraid of talking about subjective knowledge (and anecdotal evidence). To them, next to the word "science," the most sacred term may very well be "statistical test." (Beardsley, 1980, p. 75).

In this section we will try to point out that there are numerous examples which show that human knowledge is not acquired by objective methods, let alone so-called statistical estimation. For example, how does a journal editor decide to accept or reject a research paper in mathematics? Can he do so by conducting a hypothesis testing on a batch of "objective" data? For another example, how can a teacher be objective when he or she goes into a classroom? Copy everything from the book to the blackboard? For yet another example, how does a scientist select a research topic? Take a simple random sample from all papers that have ever been published? Examples like these are everywhere if one is willing to open up his mind.

It is interesting to note that although "objective" researchers do not believe subjective knowledge, they do believe *in* mathematical formulas and statistical procedures. Hence our next task is to use mathematical formulas to prove that often intuition and subjective knowledge override a 100% objective estimation. Readers who are application-oriented may skip the mathematical details in Example 1.

EXAMPLE 1   Flip a coin 100 times. Assume that 99 heads are obtained. If you ask a statistician, the response is likely to be: "It is a biased coin." But if you ask a probabilist, he may say: "Wooow, what a rare event!"[9]

My students usually push me to take a stand on this issue. They are accustomed to cut-and-dried solutions, just like certain simple-minded quantitative researchers. Here is my answer: If the coin came from K-Mart, I would tend to stick with the belief that the coin is unbiased. But if the coin came from Las Vegas, then it is a different story.

As a matter of fact, a Bayesian assessment indicates that the coin (from K-Mart) is only moderately biased. A formal analysis proceeds as follows. Let p be the probability that the coin lands heads, and y be the number of heads in n tosses. It is reasonable to assume that the prior of p is a symmetric Beta distribution with parameter a. The Beta densities with $a = 20$ and 85 are shown in Fig. 3. Note that if a is very large, then the prior will concentrate near .5. The mathematical form of the Beta density is

$$f(p) \propto p^{a-1}(1 - p)^{a-1}.$$

Hence,

$$E(p) = .5,$$
$$Var(p) = 1/\{4(2a + 1)\}.$$

We further assume that most coins from the U. S. Mint will land heads with probability p ranging from .4 to .6. This assumption implies that the standard deviation of p is about .2/6. Therefore

$$a = 112.$$

Since the likelihood function of a coin tossing is binomial, standard calculation

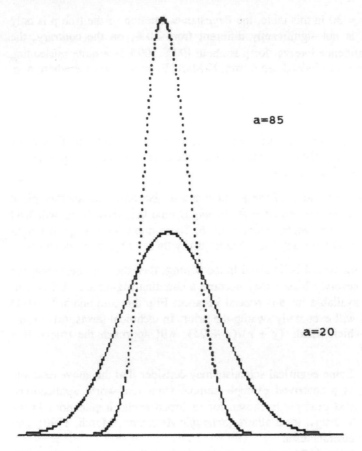

Figure 3 Symmetric beta densities with parameters a = 20 and 85.

of the posterior of p yields another Beta density with parameters (y + a) and (n − y + a). Hence

$$E(p|y) = (y + a)/(n + 2a), \tag{6}$$

$$Var(p|y) = E(p|y)(n − y + a)/(n + 2a)(n + 2a + 1). \tag{7}$$

Some applications of the formulas (6) and (7) are given in the following table.

n	y	y/n (%)	E(p|y) (%)	SD(p|y) (%)
30	29	96.7	55.5	3.1
40	39	97.5	57.2	3
50	49	98	59	3
100	99	99	65	2.6

For the case n = 30 in this table, the Bayesian estimation of the true p is only 55.5%, which is not significantly different from 50%; on the contrary, the naive 95% confidence interval for p is about [90%,100%]—a quite misleading conclusion if the coin indeed came from K-Mart. Even for the case where n = 100 and y = 99 (or the extreme case where y = n = 100), the Bayesian assessment of the coin does not support the conclusion that the coin is severely biased.

Remark 1   In this example, the coin-tossing is objective, but the Beta prior density is *not*—it was chosen mainly for the sake of mathematical convenience. Other densities may do as well.

Remark 2   The calculation of the parameter a is also based on another piece of knowledge that is not objective at all: we assume that most coins will land heads with probability ranging from .4 to .6. These numbers were pulled right off my head; I don't have any hard data to justify these "objective" numbers.

Remark 3   If big capital is involved in the betting, then the possibility that the coin is indeed severely biased may become a haunting nightmare. A frequentist solution is available for this special instance: Flip the coin another 10,000 times; the data will eventually swamp the prior. In technical terms, this means that E(p|y), which equals (y + a)/(n + 2a), will approach the true p if n goes to infinity.

EXAMPLE 2   Some empirical scientists may consider that the above example of coin-tossing is a contrived example remote from real-world applications. Therefore, the next example is chosen for its broad social implications in the testing for AIDS, drugs, legal affairs (*Scientific American*, March, 1990), and polygraph (lie detector) tests.

Assume that in an AIDS screening program, the reliability of a medical test is 95%. That is, if a person has AIDS, the test will show positive results 95% of time; if a person does not have AIDS, then the test will be negative 95% of time. Now suppose a test on Johnny is positive, what is the chance that Johnny has AIDS?

By common sense the probability that Johnny has AIDS is 95%. But people with good intuition and high respect for humanity may argue against the total reliance on the objective medical test: Assume that there are 20,000 students in a big university, and that there are 40 students who have AIDS. The error rate of medical testing is 5%, so (20,000 − 40) × 5% = 998. This means that 998 students could be wrongly identified as having AIDS, which is equivalent to a death sentence. How can "objective science" be so crude?

Such arguments have been used to combat the promotion of mass testings for drug use and AIDS (e.g., the editorial board of *The New York Times*, November 30, 1987). In most social issues, the disputes can only be settled by a vote. But fortunately, in this case a mathematical analysis can be enlisted to resolve the controversy.

Specifically, let's consider the probability tree in Fig. 4, where Pr[AIDS] = a, Pr[No AIDS] = 1 − a, Pr[Positive|AIDS] = .95, and Pr[Negative|No AIDS] = .95. By using formulas (1) and (2) in Section II, we can derive the following Bayes formula:

$$\Pr[\text{AIDS}|\text{Positive}] = \frac{.95a}{.95a + .05(1-a)}, \tag{8}$$

which is the probability that Johnny has AIDS, given the information that his test result is positive. Note that if a = 1/2, then Pr[AIDS|Positive] = .95. In other words, if a = 50%, then the common-sense conclusion that P[Johnny has AIDS] = 95% is correct. However, if a is not .5, then the answer can be quite different. For example, if a = 1/10,000 (i.e., in the whole population, out of 10,000 people, there is only one AIDS patient), then by formula (8), P[AIDS|Positive] = 0.2%. That is, the chance Johnny has AIDS is only 0.2%, not 95%. Even if Johnny is from a risk group where a = 2/1,000, the probability calculated from formula (8) is only 3.7%, not 95%.

At this moment mathematics appears to support the argument against mass testing. However, this appearance will evaporate upon close examination. Note that formula (8) can be applied recursively. For instance, if a = 1/10,000, and the medical test is positive, then by applying formula (8) once, P[Johnny has AIDS] is about 0.2%. Now if an independent test on Johnny is still positive, then we can use .2% as the new prior probability (i.e., a) and formula (8) will produce a probability of 3.7%. Repeating the same calculation 3 more times, we can conclude that P[Johnny has AIDS, given 5 positive tests] = 99.6%.

This is another example showing that repeated experiments overwhelm the prior. But repeated experiments are, in most cases, not available in soft sciences; therefore one should always double-check calculations against one's instincts. And if the two collide, then one should shelve the calculation. That is, go with your instinct, or wait for more evidence.

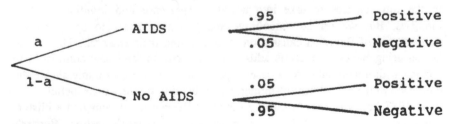

Figure 4   Tree diagram for the AIDS problem.

• • • • • • • • • •

A characteristic of the current academic world is that if a theorem has found an application, then the theorem will find its way into almost all branches of scientific disciplines. This characteristic is no exception to the Bayes formula and its extensions. One use of the Bayes formula is to justify "scientific philosophy" (Reichenbach, 1951, p. 233). The irony of this justification is that according to this philosophy, only "objective things" are real (Reichenbach, 1951, p. 261). But Bayesians who ponder the nature of probability almost unanimously deny the existence of "objective probability" (see, e.g., Good, 1988, p. 410).

Also, in sharp contrast to the "scientific philosophers," Bayesians encourage the use of subjective probability "even when estimating a physical probability" (Good, 1988). Scientific knowledge is, after all, a guide for decision making. But "when trying to make a decision," James Berger (a prominent Bayesian, 1980, p. 58) declared, "asking for objectivity is obviously silly. We should use all the information at our disposal to make the best decision, even if the information is very personal."

To Bayesians, not only information, but even the philosophical foundation of Bayesianism is personal. According to Professor Good, there are at least 46,656 varieties of Bayesians. Among them, it is doubtful that they all consent to the existence of "objective philosophy."

The textbook definition of probability is "the limit of the relative frequency of repeated events." To Bayesians, this definition is inadequate. For instance, how are we going to apply this definition to events such as "the probability there will be a nuclear war?" We certainly cannot afford to repeat this event infinitely many times. .

The interpretation of probability as the relative frequency *in the long run* is useless and self-defeating in many social-science studies. This fact was well summarized by Keynes: "In the long run, we are all dead." Bayesian theory that does not need the frequency justification of probability may provide new tools for incorporating prior knowledge and anecdotal evidence into statistical inference.

However, let the reader be cautioned. In general we encourage researchers in the soft sciences to take into account experience and intuition in their inquiries. But we do not encourage inexperienced users to apply formal mathematics of Bayesian calculation in conducting their research. The reason is that using Bayesian analysis adds a new dimension (and new confusion) in dealing with uncertainty. As a case in point, the following example shows the confusion in the assignment of numerical probability to one's prior belief.

In 1987 James Berger (a leading figure of Bayesianism) launched a blistering attack on the use of P-values in statistical hypothesis testing. Berger's papers were published in *Statistical Science* (1987a, with seven discussants) and in *JASA* (1987b, with eight discussants). The prominent status of the discussants (and the journals involved) made the discussion an unforgettable event in statistical community.

Berger's main thesis is that P-values overstate the evidence against the null hypothesis. In a zealous attempt to discredit the current practice, his quest is thus to "pound nails into the coffin of P-values." (Berger, 1987b, p. 135, *JASA*). This kind of statement is about the strongest language one can expect in scholarly publications.

Berger's argument is that Bayesian evidence against a null hypothesis can differ by an order of magnitude from P-values (1987b, p. 112):

> For instance, data that yield a P-value of .05, when testing a normal mean, result in a posterior probability of the null at *at least* .30 for *any* objective prior distribution. ("Objective" here means that equal prior weight is given the two hypotheses and that the prior is symmetric and nonincreasing away from the null.) [emphasis original]

As a good mathematician, Berger gave a solid justification to his claim (using his "objective" priors). But upon further reflection, one may be curious why equal probabilities are assigned to both hypotheses. Note that the null hypothesis in Berger's derivation is

$$H_o : \theta = \theta_o,$$

which represents a single point; while the alternative hypothesis is

$$H_1 : \theta \neq \theta_o,$$

which represents two half lines containing infinitely many points. This assignment of 50% probability to a point null reminds us of a mistake committed frequently by freshman students in introductory statistics courses. For example, you ask the students: "What is the probability a fair coin will land heads?" They will answer: "50%." So far this answer is correct. But why? Many students will say that it is because there are *two* possible outcomes, so the probability for each to happen ought to be 50%. But this argument is simply not true. Further, if you ask them:

> Flip a fair coin 100 times. What is the probability of getting 50 heads?

Most students would say 50%. But binomial formula (or a normal approximation) gives a probability of only 8%. If the number of tosses approaches infinity, then the probability it will land *exactly* half heads and half tails will approach zero.

If we apply these elementary arguments to the testing of parameters in normal or binomial distributions, we cannot accept the probability of a point null to be 50%. But this is precisely what Berger has used as one of the premises in his mathematical derivations.[10]

Berger's conclusion reveals a typical confusion of "statistical hypothesis" and "scientific hypothesis." A statistical hypothesis describes a distinct feature of a population that is narrowly defined—it is either defined for a population at a specific time (e.g., sample survey), or for a highly focused situation (e.g., repeated measurements of microwave in a very small piece of the sky). This is its limitation, but also its strength.

A scientific hypothesis, on the other hand, is something like Darwin's theory of evolution or the Big Bang theory of our universe. This kind of hypothesis does carry weight (or probability) for our belief. But a scientific hypothesis is in most cases *not* tested by formal statistical inference. For example, why does the scientific community *accept* evolution theory but *reject* the religious creation theory? By setting up null and alternate hypotheses, and then conducting a Neyman-Pearson (or Bayesian) hypothesis testing?

IV. BAYESIAN TIME-SERIES ANALYSIS AND E. T. (EXTRA TIME-SERIES) JUDGMENT

There are two different kinds of Bayesian statisticians: subjective Bayesians and objective Bayesians. In this section we will first examine the content of an "objective" Bayesian analysis in business forecast. We will then compare the "scientific" philosophy of such "objective Bayesians" to other schools of statistical analysis.

In the late 1970s, a new type of Bayesian model for economic forecasting was developed in the Research Department at the Federal Reserve Bank of Minneapolis (Litterman, 1986a, 1986b, *J. of Business and Economic Statistics*). The performance of these new models was very impressive (see Figs. 5 and 6), when compared to three professional forecasts (Chase Econometric Associates, Wharton Econometric Forecasting Associates, and Data Resource, Inc.): Both figures showed that the traditional forecasts deviated from Litterman's, while the realized values were remarkably close to the predictions made by the new method. The model that generated the previous forecast is based on a time-series technique called vector autoregressive modeling, which has been around for many years. The novelty of the new models is that the investigators (see, e.g., Litterman, 1986a, 1986b) assign a Gaussian prior distribution to the autoregressive coefficients.

The modeler (Litterman, 1986b) emphasizes that his prior distribution is not derived from any particular economic theory. The reason is that the modeler does not find "a good deal of consensus on the economic structures involved." Instead, the modeler decides to assign uniform prior (i.e., same standard deviation) to the first lag of the dependent variables in the equations. He also assigns decreased standard deviations of further coefficients in the lag distributions in a harmonic manner (Litterman, 1986b, p. 30). Forecasts are generated mechanically from the resulting Bayesian models.

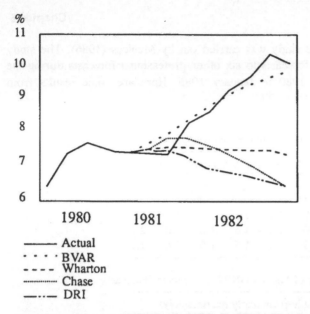

Figure 5 Comparison of unemployment rate: Forecasts as of 1981. Sources: Actual—U.S. Dept. of Labor and Commerce; Commercial—Conference Board. Copyright © 1986 by the American Statistical Association. Reprinted by permission.

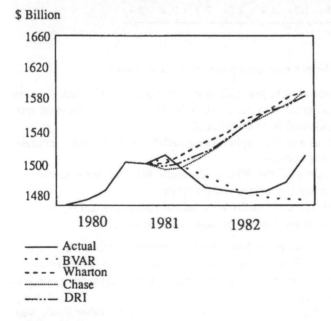

Figure 6 Comparison of Real GNP: Forecasts as of Second Quarter 1981. Sources: Actual—U.S. Dept. of Labor and Commerce, Commercial—Conference Board. Copyright © 1986 by the American Statistical Association. Reprinted by permission.

Another comparative study was carried out by McNees (1986). The study contrasts the Bayesian forecasts to six other professional forecasts during the period from February 1980 to January 1985. Here are some results from Tables 7 and 1 of McNees (1986):

Root Mean Squared Errors of Unemployment Rate Forecasts

Forecaster	Forecast horizon (early quarters only)							
	1	2	3	4	5	6	7	8
Bayesian	.4	.6	.9	1.1	1.2	1.2	1.3	1.5
Chase	.4	.7	1.0	1.3	1.6	1.8	2.0	2.1
Wharton	.3	.6	1.0	1.3	1.5	1.6	1.8	1.9

Root Mean Squared Errors of Implicit GNP Price Deflator Forecasts

Forecaster	Forecast horizon (early quarters only)							
	1	2	3	4	5	6	7	8
Bayesian	2.1	2.4	2.9	3.3	3.7	4.0	4.1	4.1
Chase	1.4	1.4	1.6	1.7	2.0	2.1	2.3	2.4
Wharton	1.7	1.5	1.7	1.9	2.0	2.1	2.3	2.4

The summary of McNees's extensive comparison is as follows.

> The Bayesian forecasts were generally the most accurate or among the most accurate for real gross national product (GNP), the unemployment rate, and real nonresidential fixed investment.
> The Bayesian forecasts of the implicit price deflator were the least accurate, especially for long horizons.
> The Bayesian forecasts of the 90-day Treasury Bill rate were among the least accurate, though by a very small margin.
> The Bayesian GNP forecast results were mixed—generally among the least accurate short term and among the most accurate long term.

McNees concluded that Bayesian forecasts can present a strong challenge to conventional forecasts and serve as a powerful standard of comparison for other forecasts. This is indeed a great contribution of statistical models to non-experimental sciences.

Litterman (the developer of the Bayesian models), on the other hand, went very far to emphasize the "objectivity" of the models and contended that conventional econometric methods rely significantly on human adjustment that "essentially takes them [the conventional models] out of the realm of *scientific study*" (1986b, emphasis mine). Bayesian models, according to Litterman, "forecast well without judgmental adjustment;" thus they can be used "to estimate *objectively* the probabilities of future events" (1986a, emphasis mine). Litterman maintained that "the most important" distinct advantage of the Bayesian approach over conventional models is that the Bayesian approach "does not require judgmental adjustment." Thus "it is a *scientific method* that can be evaluated on its own" (1986b, emphasis mine).

This stance on "scientific method" is not likely to be shared by most scholars of the philosophy of science, nor by certain researchers in the field of economic forecast. McNees (1986), for instance, contends that it is unclear why the adjustment of a mechanically generated model-based forecast should be regarded as inherently "unscientific" or necessarily more "subjective" than the process in which macroeconomic models are specified and estimated. McNees observed that the proponents of Bayesian time-series models have charged that judgmental adjustments are "unscientific" and "subjective." McNees wrote,

This charge seems to imply that scientists do not employ judgment in both the construction of their tools (models and theories) and their application. The refusal to adjust a mechanically generated forecast seems to amount to the assertion that all information of predictive value can be found in the specific sample period data used to fit the model *and* that the model has extracted all such information from this data set.

For the complicated task of modeling economy, McNees further asked,

Are all relevant factors quantifiable?
Do all future events that affect the economy have precedents in "the" sample period data?

McNees concludes that: "In my opinion, model users should be encouraged to think of models as tools that can aid in informing their judgment, rather than as substitutes for judgment that can best be employed mechanically."

A common adjustment in business forecast is related to the Federal Reserve's view of the economy and its power to manipulate interest rates, which in turn affect the value of the securities the forecasters' clients trade. Forecasters thus adjust their own evaluations of each new set of statistics (released by Government) according to how they think the Fed will interpret the data.

Is this adjustment less scientific than a mechanical procedure? Also, will the adjustment improve the mechanical Bayesian forecasts generated by Litterman's model? We believe these questions should be left to the reader's own judgment (especially to those who are going to bet their life savings in investment).

Economic forecast is seldom an easy task. In a *New York Times* news analysis (September 1, 1989), Uchitelle wrote, "Just one month ago, most economic forecasters employed by banks and brokerage houses were warning of an imminent recession; now they are celebrating continued growth. Economic miracle? Hardly." Uchitelle wrote, "Such sudden shifts have more to do with the nature of a forecaster's job than with what is actually happening to the economy."

There are plenty of reasons why business forecast is so difficult. One of them is, according to commercial forecasters, the so-called "bad numbers," which refer to the 30 or so government statistics released each month. Such statistics are often incomplete and later revised. For instance, the August 1989 statistics showed unexpected strength in employment and retail sales—contradictory to an earlier string of numbers indicating weakness in the economy. The new statistics spurred a swing in the forecast from mild recession to continued growth. Explaining the cause of this swing, Michael Penjer (a senior economist at the Bank of America) said, "We are making forecasts with bad numbers, but the bad numbers are all we've got" (*The New York Times*, September 1, 1989).

Some forecasters thrive in this volatile environment. A key to success in the profession of economic forecast is "conviction." That is, in front of the clients, forecasters have to be as firm as stone.

In comparison, there are a lot of economists "back at university jobs or driving trucks" because they "did not have the guts to stick their necks out and make a forecast" (a la Uchitelle).

•   •   •   •   •   •   •   •   •

In non-experimental (or semi-experimental) sciences, forecasting is not an easy task. We have seen some examples from social-economic science. The next example is taken from natural science.

In summer 1988, a sustained heat wave and drought provoked concern among the public, legislators, and scientists about the global warming of our planet. The issue is widely known as the "greenhouse effect," the feared heating of the earth's atmosphere by the burning of coal and oil.

In *Scientific American*, ecologists Houghton and Woodwell (1989) compared historical data on the global temperature change and on levels of heat-trapping gases. Based on such time-series data, they conclude that the world is already in the grip of a warming trend and that "radical steps must be taken to halt any further changes." In response to such "scientific" messages, certain members of congress recommend costly measures such as developing safer nuclear power plants, taxing gasoline, and increasing the fuel efficiency of automobiles.

In sharp contrast, Solow (a statistician at Woods Hole Oceanographic Institution in Massachusetts) expressed his dissent in an article entitled "Pseudo-Scientific Hot Air" (*The New York Times*, December 28, 1988). Solow observed that "the current warming started before the greenhouse effect could have begun" and that "the current warming is consistent with a mild postglacial period, probably the aftermath of the so-called 'little ice age' that ended during the 19th century." Solow's conclusion is that "the existing data show no evidence of the greenhouse effect."

A driving force behind the public and congressional attention to the issue is a NASA climatologist, James Hansen (Director of the Goldard Institute). Hansen testified to a Senate subcommittee that he was *99% certain* that heat wave was due to the buildup of greenhouse gases in the atmosphere and not due to natural variations in climate (the *Trenton Times*, February 22, 1989). Dr. Hansen's "99% probability" was calculated from a computer projection of climatic change. Such computer projections, according to Solow, are based on models (i.e., large systems of equations) representing our understanding of climate processes. But since our understanding of earth climate is limited, the models are thus of limited use. For example, these models have a hard time reproducing current climate from current data. They cannot be expected to predict future climate with any precision (Solow, 1988).

In contrast to Dr. Hansen, another NASA scientist, Albert Arking (Head of the Climate and Radiation Branch; *The New York Times*, May 23, 1989), observed that global "temperatures were decreasing or unvarying between 1940 and the late 1970's—although this period was one of strong growth in world energy consumption and fossil-fuel burning." Arking also observed that "the 50 years prior to that—from 1890 to 1940—was a period of significantly less fossil-fuel burning, yet the Earth warmed up by more than 1 degree Fahrenheit. That represents about twice the amount of the recent warming." Arking's conclusion is that it is too early to judge the relative importance of the increased greenhouse gases and it is premature to draw predictions about the future.

It may be bewildering to the public (and some academicians) that working scientists do not agree. But in any branch of non-experimental sciences, this is a norm, not an exception. In such disciplines, quoting statistics such as "99% probability" for a controversial event will hardly make it so. Dr. Hansen may be lucky that the climate turns out the way he predicts, but there is no compelling reason why his computer model is more scientific than others.

When the scientific community is divided on certain issues related to public interest, what kind of action should policy-makers take? One way is more research and more debates. Another way is by taking a poll. If after a thorough debate, a majority of people prefer living with the greenhouse effect to new taxes on gasoline (and nuclear power plants), then no action should be taken (but the voters have to bear the responsibility of their own decision). As a matter of fact, in a recent event, residents in Sacramento, California voted 53.4% to 46.6% in favor of the shut down of a publicly owned nuclear power plant (*The New York Times*, June 8, 1989). If that is what voters want, that should be what they get.

In summary, there are limitations in any branch of sciences; and when scientists are uncertain about natural laws, men have to rely on human laws. To "objective" scientists, this may be wrong. But people do have the right to be wrong.

V. A PURSUIT OF INFORMATION BEYOND THE DATA IN RANDOMIZED AND NONRANDOMIZED STUDIES

Outside the arena of time-series forecasting, "subjective judgment vs. objective estimation" is also a continuing source of scholarly exchanges. For instance, in a critique of the promotion of so-called "intelligent statistical software and statistical expert system," Velleman (1985) maintains that a statistical expert system should serve "as an aide or a colleague to the data analyst" rather than an independent "consultant" running everything by itself. The reason is that "good data analysis rarely follows a simple path." Velleman notes that often a scientist learns "more from the *process* of data analysis than from the resulting model." He further proposes some imperatives to data analysts:

Recognize the need for expert subjective judgment. What makes subjective sense will often be preferred to what fits best.

Recognize the need for real-world knowledge beyond the contents of the data.

No black box that takes data and problem description as input and produces an analysis as output can fully promote the understanding that is the true goal of most data analysis.

Velleman concludes that "the fundamental task of good science is not answering questions, but formulating the correct questions to ask" and that an "intelligent" statistical system running by itself lacks the real-world common-sense knowledge to raise many important questions.

Such appreciation of subjective knowledge and E. S. (extra statistical) judgment has long been ignored (or pushed aside) by "objective" Bayesians and certain quantitative social-behavioral scientists. In any event, the objectivity is certainly a virtue that should be treasured. But such researchers should be reminded that the use of statistical inference is inevitably subjective. For example, in regression analysis, the decision to include a variable is not all that objective. For another example, a university rejects students whose combined SAT scores are below 1200 points. The procedure, usually executed by a computer, is 100% objective. But why not 1300 or 1100 points? Is the decision to choose 1200 points as the cut-off all that objective? In addition, does the SAT really measure what it ought to measure? Will a composite of several measurements be more reliable than the "objective" SAT score?

Questions like these can go on forever. As a matter of fact, to counterbalance the dominant and narrow-minded view of objectivity in statistical inference, subjective Bayesians have long advocated the following unpopular stance:

> that decision-making is always subjective (Good, 1988)
>
> that there is much concealed subjectivity in the so-called "objective" Neyman-Pearson hypothesis testing (Diaconis, 1985)
>
> that "chance mechanisms" are part of "personal measures of uncertainty" (Dempster, 1986/1988; Good, 1987)
>
> that "the Fisher randomization test is not logically viable" (Basu, 1980), and
>
> that [objective] probability does not exist (de Finetti, 1974)—it's all in your mind.

To sum up, some prominent Bayesians tried to restore the value of (and respect for) subjective knowledge and thus went to great lengths to deny the existence of "objective probability." General scientists may feel that subjective Bayesians deny too much. But at the very least, scientists should learn this from the radical Bayesians: a curiosity of the prior and a relentless pursuit of information beyond the data.

By contrast, some statistics users (statisticians included) hold the deep conviction that we should *let data speak for themselves.*[11] This conviction is the kernel of naive empiricism that data collection and statistical analysis will automatically lead to scientific discoveries.

However, as well recognized by philosophers, facts are theory laden. For instance, Popper (1965, p. 46) wrote:

> The belief that we can start with pure observations alone, without anything in the nature of a theory, is absurd.

This biting statement always stirs violent debates among my friends and students. Some of them simply conclude that Popper's statement itself is absurd. To this interesting issue, let's compare what Poincaré (1952/1905, p. 143) had written: "It is often said that experiments should be made without preconceived ideas. That is impossible." For curve-fitting, Poincaré (p. 146) wrote: "We draw a continuous line as regularly as possible between the points given by observation. Why do we avoid angular points and inflexions that are too sharp? Why do we not make our curve describe the most capricious zigzags? It is because we know beforehand, or think we know, that the law we have to express cannot be so complicated as all that."

It is hard to describe the details of the heated debates between those for and against Popper's statement. But many who opposed Popper turned silent after they realized that theory and observation are like the question of which comes first, the chicken or the egg, and that observations are always selective. In other words, science does not begin with the gathering of data. Rather, it begins with the belief that certain data should be collected in the first place.

Furthermore, Freedman (1985) observed that good measurements are critical to good science, and to make good measurements one needs good theory.

Here is another example that would further support Popper's position. The example involves Dirac's theory and the discovery of the anti-electron as discussed in the introduction to Chapter 6. Before Dirac put forth his theory of anti-matters, physicists had studied the tracks of subatomic particles for more than 20 years. But the tracks of anti-matter had always been dismissed as statistical aberrations or things of that nature. It was Dirac's theory that guided physicists to look at our universe from a totally different perspective (Gribbin, 1984).

By comparison, soft sciences usually lack good theories; research workers therefore resort to data gathering and statistical manipulations as the goal of science. But a safe bet is that, without a good theory, even if the data are thrown in the face of a "data analyst," the chance that he will find anything significant is next to zero.

In certain disciplines the reality is that measurements are rough, theories are primitive, and randomization or controlled experiments are impossible. In such cases, what can one do (other than data gathering and statistical manipulation)? This is a question frequently asked by soft scientists.

A quick answer to this question is indeed another question: "Why don't you use your common sense?"

EXAMPLE 1   In a University lecture, a speaker tried to use time-series analysis to model the signal shown in Fig. 7.

A univariate ARIMA model certainly won't do; so he introduced a distribution to an autoregressive parameter. A "Bayesian" model then smoothed the signal and the investigator was very happy with the result. The mathematical constructs enlisted to attack the problem (such as Kalman filter and state-space modeling) were truly impressive. But in this case, questions raised by common-sense knowledge may be more revealing than the razzle-dazzle of the mathematical manipulations. According to that speaker, the signal is an earthquake data. A good scientist would try to find out the cause of the extreme variability in the signal; but the above statistician is satisfied if the data look smoothed.[12]

EXAMPLE 2   A couple who want a son have given birth to 4 girls in a row. To the couple involved, having an extra child can be a big burden physically

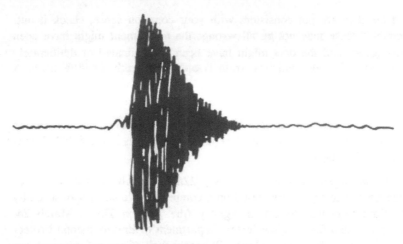

Figure 7   Time signals in a seismological study.

and financially. However, if the chance of having a son is not low (e.g., not less than 20%), then they are willing to give it a try. The issue is of much gravity to this couple. Therefore, they asked the following questions: (1) What is the probability of having 4 girls in a row? (2) What is the probability that the couple's next child will be a boy?

Textbook answers to these questions are very straightforward: (1) $.5^4 =$ 6.25%, which is not significant to reject the null hypothesis that $p = .5$; therefore the answer of (2) is 50%. And this is indeed the answer I got from a biologist.

However, common sense indicates that sex-ratio may be associated with many other factors, such as race, birth order, parental ages, etc. And this is indeed the case. (See, e.g., Erickson, 1976.) Furthermore, a physical examination of the couple might reveal that the chance their next child is a boy is one in ten million, not 50 percent.

In other words, one has to recognize (1) that statistical inference homogenizes population units and hides their differences and (2) that statistical laws do not apply to individuals. Therefore, in the process of making an important decision, one has to search for other information that may be relevant to the issue at hand. A useful technique to facilitate this process is "brain-storming" (see, e.g., Evans and Lindsay, 1989, p. 441). This technique, however, is not in the domain of statistical inference.

•   •   •   •   •   •   •   •   •   •

It is important to note that no data analysis should be accepted without a healthy dose of skepticism about what the data might have concealed. In other words, if the data are not consistent with your common sense, check it out: Your common sense may not be all wrong; the experiment might have been biased or rigged; and the data might have been contaminated or deliberately misinterpreted. After all, statistical truth is only 95% truth, or 99% truth. A measured skepticism thus may lead to useful information that was previously unnoticed.

EXAMPLE 3   A question hovers. What should you believe if your common sense conflicts with "official statistics" reported by experts in the field? Here is an example in this regard.

A federal air report ranked New Jersey 22nd nationally in the release of toxic chemicals. The report was based on a comprehensive study sponsored by the U.S. Environmental Protection Agency (the *Trenton Times*, March 24, 1989). A spokesman for the New Jersey Department of Environmental Protection was very happy about the ranking: "It speaks well of our enforcement, we feel." But New Jersey residents who have driven through northern New Jersey and gotten choked may not feel so.

As a matter of fact, the statistics in this example didn't really lie, they only got misinterpreted: New Jersey releases 38.6 million pounds of toxic chemicals annually, well behind many other states. But New Jersey is a small state, and its ranking jumped from 22nd to 4th when the tonnage release was divided by square mileage.

EXAMPLE 4   In theory, if one has a random sample from a population, then one can draw "scientific" inference based on standard formulas (assuming that no bias has crept into the sampling procedure). On the surface, the information in a random sample is "complete," and the estimated values of population parameters are thus "objective," "deductive," and "scientific." However, a pursuit of information beyond the randomized data may be very revealing.

For example, in a clinical trial, the investigator conducts a randomized experiment, but finds that there are 35 males (out of 100 male patients) in the treatment group, while the remaining 65 males are in the control group. Note that this is 3 SE (standard error) from what one would expect about the number of males in the experimental group. Also, with more males in the control, the study may give the investigator a *substantial edge* to "prove" the effect of the treatment. In such an awkward situation, should an honest scientist go ahead to do the experiment? Also, should a referee of a medical journal accept the randomized study as "scientific?"

Throughout this book, we have been advocating the deductive nature and the scientific advantages of randomized studies. The above discussion now appears to be pulling the rug from under this stance.

An "easy" solution to this problem is to separate the patients into groups of different gender, and do the randomization afterwards. But critics to randomization may argue that age is another confounding factor and that the investigator has to separate the groups of the patients further. In order to satisfy the critics, the investigator also has to take into account race, health condition, family background, and genetic differences, etc. A cruel reality is that there may exist an infinite number of confounding factors. And pretty soon one would run out of patients to perform a "valid" experiment.

But fortunately, Mother Nature has been kind to the investigators: there are only a few factors that are important, and the rest would not affect the outcome significantly. That is, only few factors are vital, and the rest are trivial. This phenomenon—vital few, trivial many—is indeed omnipresent and is one of the reasons why statistical method seems to work: If the experimental design captures the effects of major factors, then randomization will lead to the conclusion whether the treatment is beneficial.

However, findings from such experiments are still incomplete in the following sense. First, the legitimacy of the randomized experiment rests on the law of large numbers;[13] i.e., the findings will have credibility only if they are consistent with independent experiments. Otherwise, the earlier results may have to be re-evaluated.[14] Second, the experiment by itself usually cannot determine the side effects and optimum uses of the treatment.

For instance, a recent study concluded that new medicines approved by the Food and Drug Administration (FDA) caused serious reactions that required changes in warning labels or in a few cases withdrawal from the market (*Star-Ledger*, Newark, New Jersey, May 28, 1990).

The study was conducted by the General Accounting Office (GAO), a congressional investigating agency, which reported that 51% or 102 of 198 drugs studied had "serious post-approval risks" that often were not detected or disclosed by the FDA until several years after the drugs were on the market. Such unsuspected adverse reactions included heart failure, birth defects, life-threatening breathing difficulty, kidney and liver failure, blindness, and severe blood disorders, etc.

In response to the GAO report, the spokesman of the Pharmaceutical Manufacturers Association said that the report did not address the benefits of using the drugs (including those with potential risks), and that some of the drugs identified by the GAO are either the only available treatment or the best available treatment.

The U. S. Department of Health and Human Services also took issue with the GAO report. The department, in a formal reply to the GAO, objected to the methodology and the soundness of the study and expressed concern that it will "unnecessarily alarm consumers, causing some to reject the use of life-saving drugs out of fear of adverse events that might occur only in extremely rare instances."

The general public may feel uncomfortable with the disagreement among scientists. But the fact is that such episodes occur from time to time. Science, after all, does not claim certainty. Rather, it is an intellectual pursuit for a better understanding and better description of nature. As history has witnessed, when scientists think they understand everything, nature might very well give us a few more surprises. Constant challenges to the scientific establishment, if well-grounded, are thus the true spirit of science and will benefit us all.

VI. WOMEN AND LOVE: A CASE STUDY IN QUALITATIVE/QUANTITATIVE ANALYSIS

In 1972 a behavioral researcher, Shere Hite, initiated a large-scale study of female sexuality. The result was a monumental work considered by many the third great landmark in sex research, after *The Kinsey Report* (1948, 1953) and *Human Sexual Responses* (1966) by Masters and Johnson. In this section, we will examine scholarly debates over Hite's work. In addition, we will discuss the weakness of a common practice (used in Hite's work) that compares statistical breakdowns between the sample and the target population.

A motivation behind Hite's project on female sexuality was that (Hite, 1976):

> Women have never been asked how they felt about sex. Researchers, looking for statistical "norms," have asked all the wrong questions for all the wrong reasons—and all too often wound up *telling* women how they should feel rather than *asking* them how they do feel.

> The purpose of this project is to let women define their own sexuality—instead of doctors or other (usually male) authorities. Women are the real experts on their sexuality; they know how they feel and what they experience, without needing anyone to tell them.

For this purpose, Hite believed that:

> a multiple-choice questionnaire was out of the question, because it would have implied preconceived categories of response, and thus, in a sense, would also have "told" the respondent that what the "allowable" or "normal" answers would be. (Hite, 1987)

As a consequence, Hite decided to use essay questionnaires and claimed that they "are not less 'scientific' than multiple-choice." By 1976, a book based on 3,000 women's responses to her essay questionnaires was published and soon became a huge success in both sales volume and the candid revelations of how women really feel about sex.

The study was replicated (and confirmed) in at least two other countries. The book and her second volume on male sexuality have been translated into 13 languages, and are used in courses at universities in this country and around the world. Because of her work, Hite was honored with a distinguished service award from the American Association of Sex Educators, Counselors and Therapists.

An important revelation in Hite's 1976 report is that women had been compelled to hide how they feel about lack of orgasm during intercourse. A natural conclusion (Hite, 1976) is that the traditional definition of sex is sexist and culturally linked:

> Our whole society's definition of sex is sexist—sex for the overwhelming majority of people consists of foreplay, eventually followed by vaginal penetration and then by intercourse, ending eventually in male orgasm. This is a sexist definition of sex, oriented around male orgasm and the needs of reproduction. This definition is *cultural*, not biological.

In the early 1980s, Hite further sent out 100,000 questionnaires to explore how women were suffering in their love relationships with men. The response to this survey was a huge collection of anonymous letters from thousands of women disillusioned with love and marriage, and complaints about painful and infuriating attitudes on the part of men.

Hite's book, *Women and Love* (1987), thus provided a channel of release for women who experienced frequent degradation and ridicule by men. The book advocates another "cultural revolution" (instead of "sexual revolution") and was praised by some scholars:

> The Hite report on Women and Love explodes the current myth of the "insecure women" who seek out destructive relationships. . . it documents a hidden, "socially acceptable" pattern of behavior in relationships which puts women's needs last.
>
> Naomi Weisstein, Ph.D.
> Professor of Psychology
> SUNY, Buffalo

> A massive autobiography of women today.
>
> Catherine R. Stimpson
> Dean of Graduate School
> Rutgers University

> The Hite Report trilogy will long be regarded as a landmark in the literature of human behavior . . . respect will continue to grow for these works in years to come.
>
> Richard P. Halgin, Ph.D.
> Department of Psychology
> University of Massachusetts

> The most important work on men and women since Simone de Beauvoir's *The Second Sex*.
>
> Albert Ellis, Ph.D.

A distinguished report with scientific and scholarly authority.

Dr. Tore Haakinson
Wenner-Gren Center
Stockholm

By reading these reviews, one may conclude that the third Hite Report must be a masterpiece with unquestionable authority. But in fact the findings and the methodology used in the book received a thunderous rebuttal from statisticians, sociologists, public opinion pollsters, and psychologists (*Time*, October 12, 1987; *Newsweek*, November 23, 1987). In the following discussions, we will quote and discuss conflicting views on the Hite methodology.

When the first Hite Report (1976) was published, a medical writer for *The New York Times* critized the report as a "non-scientific survey." To fend off such criticisms, Hite (1987) enlisted a score of scholars to address the issue of scientific methods. Here is a representative sample of supports for the Hite methodology:

> The Hite studies . . . represent "good" science, a model for future studies of natural human experience. . . . Hite is a serious, reliable scholar.
>
> Gerald M. Phillips, Ph.D.
> Pennsylvania State University
> Editor, *Communications Quarterly Journal*

> Hite elicits from the populations she studies *not only reliable scientific data*, but also a wide spectrum of attitudes and beliefs about sex, love and who people are.
>
> Jesse Lemisch
> Professor, Department of History
> SUNY, Buffalo

> Hite has not generalized in a non-scholarly way. Many of the natural sciences worry a lot less about random samples. . . . And most of the work in the social sciences is not based on random samples either.
>
> John L. Sullivan, Ph.D.
> Professor of Political Science
> University of Minnesota
> Co-editor, *American Journal of Political Science*
> Editor, *Quantitative Applications in the Social Sciences*

It has been one of Hite's contributions to devise an excellent methodology. Hite's work has been erroneously criticized . . . as being "unscientific" because her respondents, though numerous, are not a "random sample." Hite has matched her sample carefully to the U. S. population at large; the demographic breakdown of the sample population corresponds quite closely to that of the general U.S. population.

Gladys Engel Lang
Professor of Communications,
Political Science and Sociology
University of Washington, Seattle

Hite herself was also confident about the methodology (1987, p. 777):

Sufficient effort was put into the various forms of distribution that the final statistical breakdown of those participating according to age, occupation, religion, and other variables known for the U. S. population at large in most cases quite closely mirrors that of the U. S. female population.

For instance, here are some comparisons between the study population and the U. S. female population:

Income	Study population	U. S. population	Difference
under $2,000	19 %	18.3%	+0.7%
$2,000–$4,000	12 %	13.2%	−1.2%
$4,000–$6,000	12.5%	12.2%	+0.3%
$6,000–$8,000	10 %	9.7%	+0.3%
$8,000–$10,000	7 %	7.4%	−0.4%
$10,000–$12,500	8 %	8.8%	−0.8%
$12,500–$15,000	5 %	6.2%	−1.2%
$15,000–$20,000	10 %	9.8%	+0.2%
$20,000–$25,000	8 %	6.4%	+1.6%
$25,000 and over	8.5%	8.2%	+0.3%

Type of area	Study population	U. S. population	Difference
Large city/urban	60%	62%	−2%
Rural	27%	26%	+1%
Small town	13%	12%	+1%

Geographic region	Study population	U. S. population	Difference
Northeast	21%	22%	−1%
North Central	27%	26%	+1%
South	31%	33%	−2%
West	21%	19%	+2%

Race	Study population	U. S. population	Difference
White	82.5%	83 %	−0.5%
Black	13 %	12 %	+1.0%
Hispanic	1.8%	1.5%	+0.3%
Asian	1.8%	2 %	−0.2%
Middle-Eastern	0.3%	0.5%	−0.2%
American Indian	0.9%	1 %	−0.1%

By looking at these comparisons, it is tempting to conclude that the study population represents well the U.S. population. But this is not the case, and unfortunately this is the position taken by Hite and by those who praised her work as a scientific survey.

For example, Hite was confident enough to refute both the percentage and methodology of *The Kinsey Report*:

> In 1953, Alfred Kinsey found that 26 percent of women were having sex outside of their marriages. In this study, 70 percent of women married five years or more are having sex outside their marriage. This *is* an enormous increase. Women's rate of extramarital sex has almost tripled in thirty-five years. Kinsey's figures were just rather low because he was conducting face to face interviews.... This may have inhibited reporting on the part of some interviewees. The present study is based on anonymous questionnaires, so that women would have had no reason to conceal anything.

But *Time* (1987) pointed out that "more recent studies, including the *Redbook* poll, have shown little change [in the infidelity rates]" from Kinsey's original 26%. Another random sample of 2,000 people (Patterson and Kim, 1991) shows that 31% of married people are having or have had an affair, drastically different from Hite's 70% figure.

In one chapter entitled "Ten Women Describe Their Marriages," Hite (1987, p. 346) wrote: "The women in this chapter represent the more than 2,000 married women participating in this study. Their responses were chosen because they represent well the thoughts and feelings of other married women." And her conclusion about marriage is as follows:

> Feminists have raised a cry against the many injustices of marriage—exploitation of women financially, physically, sexually, and emotionally. This outcry has been just and accurate.

However, a 1987 Harris poll of 3,000 people found that family life is a source of great satisfaction to both men and women, with 89% saying their relationship with their partner is satisfying.[15]

Some of the questions in Hite's survey appeared to solicit men-bashing. For instance, 69% of women in her study say that men seem confused by falling in love, and 87% say it is difficult to meet men they admire and respect. Such findings, although true in certain cases, are at odds with the Harris poll mentioned above.

In another instance, Hite reported that 98% of women want to make basic changes in their relationships. Well, everybody would agree that things in love relationships cannot be 100% perfect. But as warned by Tom Smith of the National Opinion Research Center: "Any question you asked that got 98% is either a wrong question or wrongly phrased" (*Time*, October 12, 1987).

Some scholars asserted that the Hite survey was designed to support her preconceived feminist notions. And perhaps the most hard-edged comment was from Ellen Goodman, a Pulitzer-prize-winning columnist: "She goes in with a prejudice and comes out with a statistic" (*Time*, October 12, 1987).

Note that the sample size in Hite's report was 4,500, a lot larger than most of the Gallup Polls. So what went wrong? For one thing, the response rate of Hite's survey is 4.5%. Just by common sense, one may wonder about the other 95.5%. Those women may not have a problem with men and thus did not bother to respond. According to chairman Donald Rubin of the Harvard statistics department (*Newsweek*, November 23, 1987), one should look for a response rate of 70 to 80 percent in order to draw valid conclusions in this type of study.

•   •   •   •   •   •   •   •   •   •

Upon further reflection, one may wonder about the reasons of those who chose not to respond. Here are some possible answers: It is hopeless to try to change the way men treat women; it is time-consuming to answer a long questionnaire; "I" am only one of millions who suffer, and it won't make any difference if "I" do not respond.

We believe that the actual number of women disillusioned with marriage and love would be a lot higher than the 4.5% who responded to the Hite survey. Assume that there are 70 million women in this country and only 10% are so unlucky as to have chosen the "wrong man," then there would be 7 million people involved in the study, more than the number of patients of any single disease in modern society.

For this reason, in this subsection, we will turn our discussion to the insight and the contribution of the Hite Reports. After all, her first report was ground-breaking and deserves a place in history. Her third report, although seriously muddled, addresses issues tied to real and pressing needs of human beings.

In the third Hite Report (*Women and Love*, 1987, p. 246), a woman wrote with self-awareness and great passion:

Love relationships with men aren't fair; but so what, I still want one. I just want to know what I have to do to get one to work.

We believe that many women share the same attitude for love. In public, every educated person would agree human beings are created equal; but in personal relationships, many women still assume (or are forced to assume) the role of "caretakers," while men remain the "taker" (Hite, 1987, p. 771). With increasing numbers of women entering the work force and sharing the financial burden of modern life, it is expected that more women must balance husband, children, and career. It is also expected that fairness and equality are only mirages in many relationships.

The third Hite Report, *Women and Love*, is in many ways a moving tale of repression and humiliation. Page after page, one is struck by the despair, alienation, and rage that certainly exist in many corners of our society. Professor Joseph Fell of Bucknell University summarized what has been revealed in Hite's books: "I find in the Hite Reports a massive cry of the human heart, a cry for the open recognition of the fundamental needs."

But such open recognition was seldom revealed in depth by the traditional methods that use random sample and multiple-choice questionnaires. This fact is complimented by scholars who are tired of dominant "objective" statistical methods (Hite, 1987):

> [Hite's survey used] questionnaires that invited long and detailed replies rather than brief, easily codable ones.... And from these responses, presented in rich, raw detail, deep truths emerge—truths which, in many cases were probably never revealed to anyone else before.
>
> Robert L. Carneiro, Ph.D.
> Curator of Anthropology,
> The American Museum of
> Natural History, New York

> Statistical representativeness is only one criterion for assessing the adequacy of empirical data... other criteria are particularly pertinent when looking at qualitative data. This is primarily the situation with Hite's research.
>
> Robert M. Emerson, Ph.D.
> Professor of Sociology, UCLA
> Editor, *Urban Life: A Journal*
> *of Ethnographic Research*

> Hite's research, as with all research dealing with thoughts and feelings, cannot be expected to be analyzed with the same techniques as those that tell us what doses of what drugs give what results in what kinds of patients.... We are a scientifically illiterate people, and honest scientists such as Hite are bound to suffer as a result.
>
> Mary S. Calderone, M.D., MPH
> Founder, Sex Education and
> Information Council of the U.S.

The Hite Reports are part of an international trend in the social sciences expressing dissatisfaction with the adequacy of simple quantitative methods as a way of exploring people's attitudes.

Jesse Lemisch
Professor, Department of History
SUNY, Buffalo

We basically agree with these scholars' support of the Hite methodology in that her methods are excellent instruments to penetrate into profound realities that had been previously unexplored. Much of the behavioral research we have come across (see, e.g., Chap. 1, Sec. I, Chap. 3, Sec. I, and Chap. 5, Sec. II) is loaded with statistical analysis but exhibits little intellectual content. Hite's approach, albeit naive in statistical methods, produced a landmark report and revealed much of the dark side of male-female relationships.

Nevertheless, we are appalled that the "Hite methodology" was used (and praised) in her statistical generalization from a sample to a population. We believe that what Hite should have done is a two-step procedure: (1) use essay questionnaires to explore depth and variety of responses, and (2) use a random-sample (and multiple-choice questions) to *confirm* statistical issues unearthed by EDA[16] (exploratory data analysis) in step (1).

This second step is called CDA (confirmatory data analysis) in statistical literature. In Hite's case, the CDA can be easily done by hiring a pollster to confirm or refute the original findings.

For instance, 76% of the single women in Hite's study say that they frequently have sex on their "first date." The opposite of this number (by which we mean 24%) would be more likely, and a random sample should be conducted to assess the accuracy of the finding.

Hite (1987) reported that "to go from essay statements to mixed quantitative/qualitative data is a long and intricate process." She estimated that "in this study, there were over 40,000 woman-hours involved in analyzing the answers, plus at least 20,000 put in by the women who answered the questionnaire. This of course does not include the time and effort needed to turn the resulting compilation of data into a book."

However, more woman-hours do not automatically lead to a more scientific conclusion, and the two-step procedure might have reduced the 40,000 hours Hite used to analyze the answers to 30,000 or less.

In sum, EDA often gives us more insights; but findings from EDA can also give us less than they seem to. Research workers who are not equipped to tell the difference between EDA and CDA may soon find themselves laughing stocks just like Hite (and scholars who praised Hite's "scientific" results).

•   •   •   •   •   •   •   •   •

An interesting fact is that Hite used the same methodology to produce three books (female sexuality, male sexuality, and women and love), but only the third book received a glare of public scrutiny (and ridicule). In her first report, Hite correctly pointed out that the traditional definition of female sexuality is *cultural*, not biological. Her findings on female orgasm, however, are *biological*, not cultural. The main point here is that many biological phenomena (such as hunger, thirst, and desire) are relatively universal. Without randomization, Hite's sample incidentally matched the population at large.

The third Hite Report (*Women and Love*), however, deals with attitudes and emotions, which are not all that universal any more. Without a random sample, generalization from a sample to a population can result in a collapse of credibility.

As an analogy to the first and the third Hite Reports, assume that you want one cup of homogenized milk and one cup of orange juice from two big bottles of both. For the cup of milk, you just pour it from the big bottle. But for the cup of orange juice, you have to shake the bottle.

Professor John Sullivan at the University of Minnesota (and editor of two academic journals) pointed out that "most of the work in the social sciences is not based on random samples." He further wrote (see Hite, 1987):

> in fact, many if not most of the articles in psychology journals are based on data from college students, and then generalized. Interestingly, they are not criticized in the same way Hite has been.

This is true. And this is why most findings in psychology journals (and other soft sciences as well) have to be looked at with caution. Otherwise, these so-called "scientific" findings may be a source of more riddles than clues.

NOTES

1. To the dismay of logical positivists, the statement is logically self-contradictory.
2. See Will, 1986, p. 65.
3. The Newtonian value of .87″ can be obtained (1) by assuming that light is a particle undergoing the same gravitational attraction as a material particle, or (2) by using the equivalence principle of relativity theory without taking into account the curvature of space.
4. Upon close inspection of the table, we find that the second data set (SD = .48, P = 7%) did not pose serious threat to Einstein's theory. Furthermore, one can calculate a weighted average of the mean deflections in the table. Specifically, let

$$G = a * \bar{X} + b * \bar{Y} + (1 - a - b) * \bar{Z}.$$

Then a standard procedure is to choose (a,b) that minimizes the variance of G. Simple calculation yields the following results:

$$a = \frac{V2 * V3}{V1 * V2 + V2 * V3 + V3 * V1,} \quad b = \frac{V1 * V3}{V1 * V2 + V2 * V3 + V3 * V1,}$$

where $V1 = Var(\overline{X})$, $V2 = Var(\overline{Y})$, and $V3 = Var(\overline{Z})$. With these choices of a and b, the weighted mean deflection is 1.815, with an SD = .1562, which is clearly in support of Einstein's theory. There is no need to throw out the second data set, as Eddington did.

5. This example was called to my attention by Peter Lakner, currently at New York University. Bradly Efron has developed a similar example that uses four dice.

6. This example was called to my attention by Moon-Chan Han, currently at the University of California, Berkeley. For further development of this example and other related issues, see Coyle and Wang (1993).

7. The spirit of the Dubins-Savage strategy is Bayesian, in that it deals with events that will not be repeated over and over again. See the subsequent sections for further discussions of Bayesian probability.

8. The two-slit experiments have never been actually performed. The results of the experiments are known by extrapolation from other real experiments.

9. This example was called to my attention by Professor David Boliver at Trenton State College, New Jersey.

10. Note that a statistical test (e.g., a P-value) provides only a quick benchmark to assess the parameter of a population. After this quick check (or without this quick check), one still has to report the confidence interval in assessing a normal mean. Berger (1987a) proposed to use both confidence interval and a Bayes factor hinged on the likelihood principle in scientific reports. This is an interesting proposal. Nevertheless, there is no compelling reason why everybody has to subscribe to the likelihood principle.

    See also Chapter 1, Section H (or Olshen, 1973) for the difficulty in the calculation of the conditional probability of simultaneous coverage, given that statistical test has rejected the null hypothesis.

11. This is exactly what I was told by a statistician in his demonstration of AUTOBJ, a software package that produces Box-Jenkins forecasts automatically. The statistician has an advanced degree from the University of Wisconsin, Madison. The guy *literally* believes that a user will be better off if he or she has no background knowledge of the data.

12. Non-mathematicians usually do not understand the utility and the esoteric beauty of mathematics. In self-defense and self-promotion, certain mathematical statisticians are in the habit of applying sophisticated techniques to data that has no connection with the techniques.

13. The rockbottom of hard sciences (such as physics and chemistry) are results that have been challenged, revised and repeated over and over again.

14. For example, it is well known that a random sample may not be a *representative* sample of the whole population. Adjustment of the results according to the census data may substantially improve the accuracy of the sample data. Examples of this type are "ratio estimation" in Gallup Polls or in the estimation of unemployment rates (Freedman, Pisani, and Purves, 1978, pp. 314, 363).

15. However, in the Patterson-Kim 1991 survey, 47% aren't sure they would marry the same person if they had to do it over again.

16. Here we include Hite's QEA (qualitative exploratory analysis) as a subcategory of EDA, although QEA involves only primitive data analysis.

REFERENCES

Accardi, L. (1982). Quantum Theory and Non-Kolmogorovian Probability. *Stochastic Processes in Quantum Theory and Statistical Physics*, edited by Albeverio, Combe, and Sirugue-Collin. Springer-Verlag, New York.

Accardi, L. (1985). Non-Kolmogorovian Probabilistic Models and Quantum Theory. Invited talk at the 45th ISI session. A preprint.

Basu, D. (1980). Randomization Analysis of Experimental Data: The Fisher Randomization Test. *JASA*, Vol. 75, No. 371, 575–582.

Beardsley, P. (1980). *Redefining Rigor*. Sage, Beverly Hills.

Bell, J.O. (1964). On the Einstein-Podolsky-Rosen Paradox. *Physics*, Vol. 1, No. 3, 195–200.

Berger, J.O. (1980) *Statistical Decision Theory: Foundations, Concepts, and Methods*. Springer-Verlag, New York.

Berger, J.O. and Delampady, M. (1987a). Testing Precise Hypotheses. *Statistical Science*, Vol. 2, No. 3, 317–352.

Berger, J.O. and Sellke, T. (1987b). Testing a Point Null Hypothesis: The Irreconcilability of P Values and Evidence. *JASA*, Vol. 82, No. 397, 112–122.

Breiman, L. (1985). Nail Finders, Edifices, and Oz. *Proceedings of the Berkeley Conference in Honor of Jerzy Neyman and Jack Kiefer*, Vol. 1, 201–214. Wadsworth, Belmont, California.

Carnap, R. (1934/1936). *The Logical Syntax of Language*, Harcourt, New York.

Coyle, C. and Wang, C.W.H. (1993). Wanna Bet? On Gambling Strategies that May or May not Work in a Casino. To appear in *The American Statistician*.

Davis, P.J. and Hersh, R. (1981). *Mathematical Experience*. Birkhauser, Boston.

deFinetti, B. (1974). *Theory of Probability*. Wiley, New York.

Dempster, A.P. (1988). Employment Discrimination and Statistical Science. *Statistical Science*, Vol. 3, No. 2, 149–195.

Diaconis, P. (1985). Bayesian Statistics as Honest Work. *Proceedings of the Berkeley Conference.in Honor of Jerzy Neyman and Jack Kiefer*, Vol. 1, 53–64. Wadsworth, Belmont, California.

Dubins, L.E. and Savage, L. (1965). *Inequalities for Stochastic Processes: How to Gamble if You Must*. Dover, Mineola, New York.

Earman, J. and Glymour, C. (1980). Relativity and Eclipses: The British Eclipse Expeditions of 1919 and their Predecessors, in *Historical Studies on the Physical Sciences*, edited by G. Heilbron et al., Vol. 11, 49–85. University of California Press, Berkeley.

Erickson, J.D. (1976). the Secondary Sex Ratio in the United States 1969-1971: Association with Race, Parental Ages, Birth Order, Paternal Education, and Legitimacy. *Ann. Hum. Genet*, Vol. 40, 205–212.

Feynman, R. (1948). Space-Time Approach to Non-Relativistic Quantum Mechanics. *Rev. of Mod. Phys.*, Vol. 20, 367–385.

Feynman, R. (1951). The Concept of Probability in Quantum Mechanics. *Pro. II-d Berkeley Symp. on Math. Stat. and Prob.*, 533–541. University of California Press, Berkeley.

Fienberg, S.E. (1986). Reply to "Statistics and the Scientific Method" by D.A. Freedman. *Cohort Analysis in Social Research*, edited by W.M. Mason and S.E. Fienberg, Springer-Verlag, New York, pp. 371–383.

Freedman, D.A. (1985). Statistics and the Scientific Method. *Cohort Analysis in Social Research*, edited by W.M. Mason and S.E. Fienberg, Springer-Verlag, New York, pp. 343–366.

Freedman, D. A. (1987). As Others See Us: A Case Study in Path Analysis. *J. of Educational Statistics*, Vol. 12, No. 2, 101–128, 206–223; with commentaries by 11 scholars, 129–205.

Good, I.J. (1987). A personal communication.

Good, I.J. (1988). The Interface between Statistics and Philosophy of Science. *Statistical Science*, Vol. 3, No. 4, 386–412.

Gribbin, J. (1984). *In Search of Schrodinger's Cat: Quantum Physics and Reality*. Bantam Books, New York.

Harris, M. (1980). *Cultural Materialism: The Struggle for a Science of Culture*. Vintage, New York.

Hite, S. (1976). *The Hite Report: A National Study on Female Sexuality*. Macmillan, New York.

Hite, S. (1987). *Women and Love: A Cultural Revolution in Progress*. Knopf, New York.

Hofstadter, D.R. (1979). *Godel, Escher, Bach: an Eternal Golden Braid*. Vintage, New York.

Kuhn, T. (1970). *The Structure of Scientific Revolutions*, 2nd edition. University of Chicago Press.

Litterman, R.B. (1986a). A Statistical Approach to Economic Forecasting. *J. of Business and Economic Statistics*, Vol. 4, No. 1, 1–4.

Litterman, R.B. (1986b). Forecasting with Bayesian Vector Autoregressions–Five Years of Experience. *J. of Business and Economic Statistics*, Vol. 4, No. 1, 25–38.

McNees, S.K. (1986). Forecasting Accuracy of Alternative Techniques: A Comparison of U.S. Macroeconomic Forecasts. *J. of Business and Economic Statistics*, Vol. 4, No. 1, 5–15.

Ohanian, H.C. (1987). *Modern Physics*. Prentice Hall, Englewood Cliffs, New Jersey.

Patterson, J. and Kim, P. (1991). *The Day America Told the Truth*. Prentice Hall, Englewood Cliffs, New Jersey.

Popper, K. (1965/1963). *Conjectures and Refutations: The Growth of Scientific Knowledge*. Harper and Row, New York.

Poincaré, H. (1952/1905). *Science and Hypothesis*. Dover, Mineola, New York.

Regis, E. (1987). *Who Got Einstein's Office? Eccentricity and Genius at the Institute for Advanced Study*. Addison Wesley, Reading, Massachusetts.

Reichenbach, H. (1951). *The Rise of Scientific Philosophy*. University of California Press, Berkeley.

Russell, B. (1948). *Human Knowledge, its Scope and its Limits*. Simon and Schuster, New York.

Russell, B. and Whitehead, A.N. (1910). *Principia Mathematica*. Cambridge University Press, Cambridge, Massachusetts.

Schlick, M. (1931). *The Future of Philosophy*. Oxford.

Shimony, A. (1988). The Reality of the Quantum World. *Scientific American*, Vol. 258, No. 1, 46–53.

Speed, T.P. (1985). Probabilistic Risk Assessment in the Nuclear Industry: Wash-1400 and Beyond. *Proceedings of the Berkeley Conference in Honor of Jerzy Neyman and Jack Kiefer*, Vol. 1, 173–200. Wadsworth, Belmont, California.

Velleman, P.F. (1985). Comment to "More Intelligent Statistical Software and Statistical Expert Systems: Future Directions" by G.J. Hahn. *American Statistician*, Vol. 39, No. 1, 10–11.

Will, C.M. (1986). *Was Einstein Right?* Basic Books, New York.

Chapter 7

A Delicate Balance Between Order and Chaos

I. OBJECTIVITY VERSUS RELIABILITY

Objective procedures have long been the essence of science. Strong proponents of objectivity can be traced back to Francis Bacon, who proposed that data gathering be carried out by illiterate assistants with no interest in whether an experiment turned out one way or another (Harris, 1980). Bacon's proposal was fortunately ignored by scientists. Nevertheless, science is commonly associated with objective knowledge, while subjective judgment is equated with wishful thinking that should have no place in scientific endeavors.

The pursuit of objectivity in scientific reports is certainly a virtue that should be treasured. But here is a note of caution: in soft science, "objectivity" does not entail "reliability," and "objective" methods may be far from "effective."

This is especially important for studies in which the measurements are rough, the theoretical foundation is weak, and statistical computations replace intellectual reasoning. Unfortunately, the belief in objective procedures among semi-experimental scientists has been running so deep that it defies easy communication. Here are some cases in point.

EXAMPLE 1 Suppose that a person sues for health damage (e.g., cancer) caused by a certain chemical product, and that direct evidence from experiments on human beings is lacking. What kind of evidence should be used in settling the law suit?

A conventional procedure involves experiments on mice, following which a dose-response model is used in numerical extrapolation from animals to humans. One dose-response model is called the one-hit model, and can be described as follows:

$$P(d) = P(O) + [1 - P(O)] * (1 - e^{-kd}),$$

where P(d) is the total chance of cancer, at dose d, due to the exposure and to all other causes; and k is the potency parameter. Freedman and Zeisel (1988) discussed the one-hit model, as well as the multi-hit and Weibull models commonly used in risk assessment.

Freedman attacked the routine risk-assessment approach on the grounds that (1) the dose-response models are far removed from biological reality; (2) different models give substantially different risk estimates; (3) generalization from animals to humans is not statistically justifiable; and (4) extrapolation from high doses to low doses is not scientifically possible.

Freedman contends that it may be advisable to give up the pretense of a scientific foundation where none exists, and that there is little hope for progress in risk assessment until the biology of cancer is better understood.

Statistics as a science is to quantify *uncertainty*, not *unknown*. But equations in risk-assessment appear to do the opposite. Given the weak foundation of the dose-response models, Freedman concluded that

> we see no evidence that regulatory modeling leads to better decisions than *informal argument*, and find the latter more appealing because it brings the uncertainty into the open. [emphasis supplied]

To some, this is a call for uprisings. Outraged by this conclusion, J. K. Haseman (a Research Mathematical Statistician at the National Institute of Environmental Health Sciences) complained that

> When scientific evaluations of possible human risk are replaced by a lawyer's debating skills, and regulatory decisions determined solely by who is able to "informally argue" their position more persuasively, then the ultimate loser may be our nation's public health. [a la Freedman]

Future studies may prove that Freedman is wrong, similar to the case of smoking and lung cancer (see Chap. III, Sec. II; or Freedman, Pisani, and Purves, 1978, p. 11). But given the weak foundation of the risk assessment, we don't see why informal arguments are less scientific than the dose-response models.

It is instructive to note that statistics used to be a branch of political science (before its full development for randomized experiments). And it may be healthier to regard statistics as a subdomain of political science or psychology, in the cases that experimental (or semi-experimental) results are stretched beyond their credential. This view of statistics may look retrogressive, but it is in accord with a great tradition of western civilization that respects *both* science and democracy. Thus, when an issue cannot be resolved by scientific methods, then we have to rely on subjective judgment, experts' opinions (although they often conflict each other), group discussion, and finally a vote. What can be more scientific?

EXAMPLE 2　In their book *Discovering Causal Structure: Artificial Intelligence, Philosophy of Science, and Statistical Modeling*, Glymour et al. (1987) quoted a study by Paul Meehl (1954) which compared predictions obtained by psychologists using clinical interviews with those obtained by simple linear regression of the predicted attribute on any statistically relevant features of the population. Meehl's study found that on the average the predictions obtained by linear regression were never worse, and were usually better, than those made by clinicians.

According to Glymour, Meehl's conclusion is one of the best replicated results in psychology. Glymour then concluded that

> if the issue is selecting police officers who will perform their task satisfactorily, or graduate students who will do well in their studies, or felons who will not commit new crimes when paroled, *a simple statistical algorithm* will generally do better than a panel of Ph.Ds. [emphasis supplied]

Glymour further observed that

> Results of this sort are not confined to clinical psychologists. A growing literature on "behavioral decision theory" has found that people, even those trained in probability and statistics, perform well below the ideal in a variety of circumstances in which judgments of probability are called for. [Kahreman, Slovic, Tversky, 1982]

Glymour's analysis appears to support the common belief that the objective approach is better than human judgment. His arguments sound very convincing, at least on the surface. But let's consider his example of the recruitment of graduate students. Many people may be able to live with the idea of using a simple statistical algorithm for the recruitment of graduate students. But how about the search for a faculty member? a Dean? the President of a large university? In such instances, the screening of the candidates is very subjective, although it may be part of a systematic procedure. And it will be absurd to replace the human judgment in these cases by a simple statistical algorithm.[1]

The irony in this example is that the search for a faculty member is much *more important* than the recruitment of a graduate student. Yet we use a lot more subjective judgment in the faculty search than in student recruitment. Does this fact ring a bell for the quantitative researchers who are so afraid of subjective judgment?

EXAMPLE 3   In the 1987 ASA (American Statistical Association) annual meeting, a conference room was packed with statisticians and econometricians who were curious about the question, "How Good Are Econometric Models?" The theme session was organized by David Freedman of the University of California at Berkeley. Scholars who presented their papers or participated in the discussion included Lincoln Moses of Stanford University, Dan McFadden of the Massachusetts Institute of Technology, Thomas Sargert of the University of Minnesota, and Stephen McNees of the Federal Reserve Bank of Boston.

The mood in the room was a healthy skepticism about the statistical models used in econometric research, and there was a sense that human judgment might give us better insight into a complicated phenomenon.

One of the speakers was against this mood. His argument was that a systematic approach is in the long run better than ad hoc treatments. This argument met no challenge from the audience; but silence does not mean consensus. Instead, it usually means that opponents are short of ammunition (and rhetoric) at that specific moment.

In principle we believe that effective *management* relies on objective and systematic methods, and that ad hoc treatment is a poor man's short-sighted solution. But is it correct to equate systematic approach with econometric research? Or is it correct to equate a management philosophy with statistical data analysis? Readers might recall the heated debates over Bayesian time-series in particular (Chap. 6, Sec. IV), and the dismal performance of econometric research in general (Chap. 4, Sec. II). It is now up to you to choose a side.

II. ORDER WITHIN CHAOS: HEARTBEATS, BRAINWAVES, SIMULATED ANNEALING, AND THE FIDDLING OF A SYSTEM

In soft science, debates on the issues of "systematic (or objective)" approach can go on forever, because many researchers have trouble with "informal arguments" that are deemed "merely" philosophical and do not have the prestige of mathematical equations. For this very reason we will present some examples from the science of chaos and dynamical systems (a new branch of mathematics). The discussions will also take us to another level of understanding.

EXAMPLE 1   Consider the following graphs of heart-rate time series (Fig. 1). The first heart-rate time series is somewhat chaotic, while the second is almost perfectly predictable. Which time series do you prefer to be associated with your own heart?

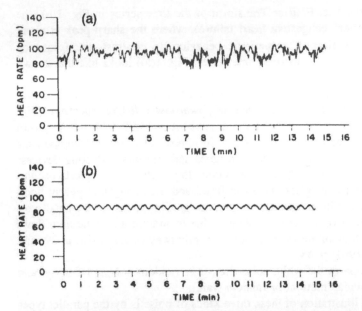

Figure 1 (a) Healthy subjects show considerable beat-to-beat activity. (b) Heart rate time series for patient with severe congestive heart failure. Copyright © 1988 by Birkhauser. Reprinted by permission.

This question was presented (without the graphs) to three biology professors (two of them have extensive experience with ECG, the electrocardiographic data). They all responded that the predictable heart-rate time series is associated with a healthy heart, while a chaotic heartbeat is likely to presage sudden death. Not true, at least not for the above times series.[2]

Goldberger and co-workers (1984, 1985, 1987, 1988, 1989, etc.) have completed a series of investigations of the nonlinear dynamics in heart failure. They found that the time intervals between beats of a healthy heart constantly vary by small amounts, and that these variations are completely unpredictable. They conclude that a healthy heart must beat somewhat chaotically rather than in a perfectly predictable pattern.

In a study of 30 dying patients, Goldberger found that a few hours before death, the intervals between beats tend to lose their natural variation and become practically identical. Goldberger concluded: "The healthy heart dances, while the dying organ can merely march" (quoted from *The New York Times*, January 17, 1989). Another interpretation is that a healthy heart responds well to a changing environment, while a dying heart loses its ability to adjust for natural variation (just like certain quantitative researchers who do everything mechanically).

Figure 2 is the Fast Fourier Transform of the time series in Fig. 1(b) (the patient had a severe congestive heart failure), where the sharp peak at about .03 Hertz is an indication of periodic behavior. In contrast, the frequency spectrum of the healthy heart-rate time series (Fig. 1(a)) looks like that shown in Fig. 3.

If one plots the logarithms of both x (frequency) and y (amplitude) in Figure 3, then one gets the so-called inverse power-law (1/f-like) spectrum (Fig. 4), where the slope of the regression line is about -1.005. This 1/f spectrum is omnipresent in such physical systems: in almost all electronic components from simple carbon resistors to vacuum tubes and all semiconducting devices; in all time standards from the most accurate atomic clocks and quartz oscillators to the ancient hour-glass; in ocean flows and the changes in yearly flood levels of the river Nile as recorded by the ancient Egyptians; in the small voltages measurable across nerve membranes due to sodium and potassium flow; and even in the flow of automobiles on an expressway (Voss, 1988, p. 39; and Bak and Chen, 1991, p. 48).

Figure 5 compares the white noise (i.e., pure random noise), 1/f noise, and $1/f^2$ noise (Brownian motion, or random walk).

An excellent illustration of these three kinds of noise is by the parallel types of computer-simulated "fractal" music (Fig. 6). These analyses indicate that nature favors a delicate harmony between order and chaos. Moreover, a reasonable measure of this harmony is the 1/f-spectrum.[3]

Further appreciation of this 1/f-like distribution can be gained from Figs. 7(a)–(c), which shows the heart-rate time series of three patients and the corresponding spectrums. A common feature among the sick hearts is the fact that their frequency spectrums cannot be fitted well to a 1/f-like distribution.

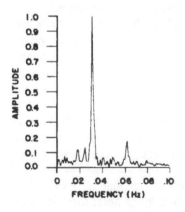

Figure 2   Oscillatory pattern with a sharp spectral peak at about 0.03 Hz. Copyright © 1988 by Birkhauser. Reprinted by permission.

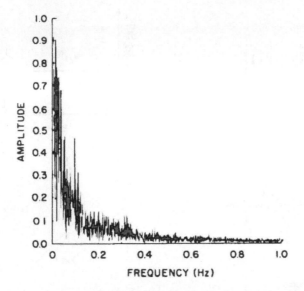

Figure 3   Frequency spectrum of the healthy heart-rate time series. Copyright © 1988 by Birkhauser. Reprinted by permission.

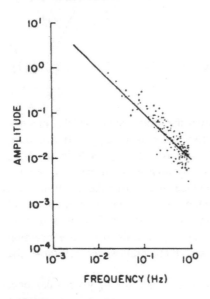

Figure 4   Inverse power law (1/f-like) spectrum.

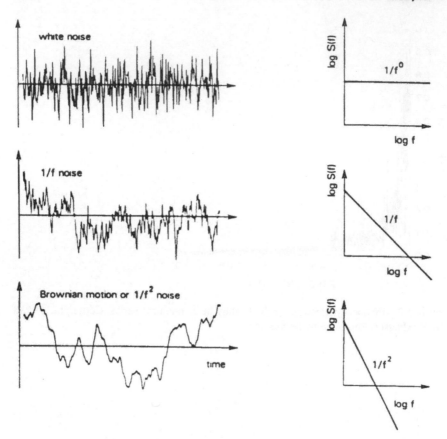

Figure 5   Comparison of white noise, 1/f noise and $1/f^2$ noise. Copyright © 1988 by Springer-Verlag. Reprinted by permission.

As a further illustration, Figure 8 shows the brain waves (EEG, electroencephalogram) of two individuals. The top wave looks ugly and irregular. The bottom one looks extremely regular: nice round wave is followed by a sharp wave, another round wave is followed by another sharp wave. Garfinkel (1989, NOVA #1603: the Strange New Science of Chaos) asked: "Which one would you rather have?" In my experience, some psychometricians would probably choose the bottom one (see Chap. 4, Sec. II, "A Religious Icon in Behavioral Research"). These researchers have been trained to do everything "objectively" and "systematically." Perhaps their brainwaves resemble the bottom, regular EEG.

Here is the answer from Garfinkel: The upper one, the ugly irregular one, is a normal adult EEG; the bottom one is a spike wave complex of an epileptic seizure, which is a very dangerous and pathological condition.

Figure 6   (a) "White" noise is too random; (b) "1/f" noise is closest to actual music (and most pleasing); (c) "Brown" music is too correlated. Copyright © 1988 by Springer-Verlag. Reprinted by permission.

•   •   •   •   •   •   •   •   •

Scientists are now using sophisticated mathematics to shed new light on phenomena that are usually called chaos. Some surprising conclusions from this new branch of science are that (1) often there is order in chaos, and (2) some control systems can function successfully only by introducing noise (a chaotic component) into the system. The first conclusion is referred to the 1/f-distribution and Gleick (1987), while the second conclusion will be illustrated by the following examples.

EXAMPLE 2   Scientists often want to minimize an objective function over a configuration space in which parameters are not continuous and elements are factorially large (so that they cannot be explored exhaustively). Traditional algorithms for finding the minimum may go from the starting point immedi-

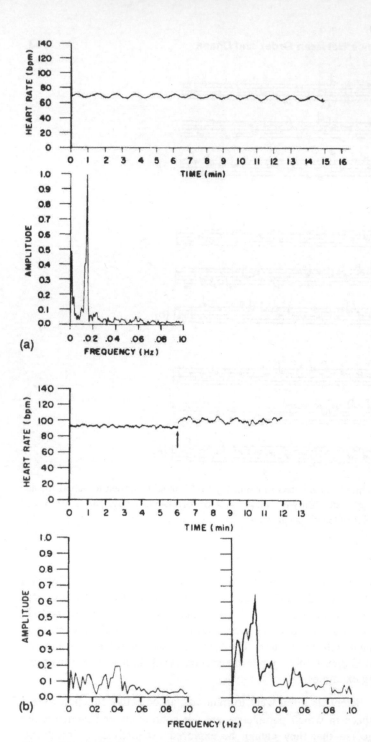

Figure 7 (a)–(c) Heart-rate time series of 3 patients and their corresponding power spectra. Copyright © 1988 by Birkhauser. Reprinted by permission.

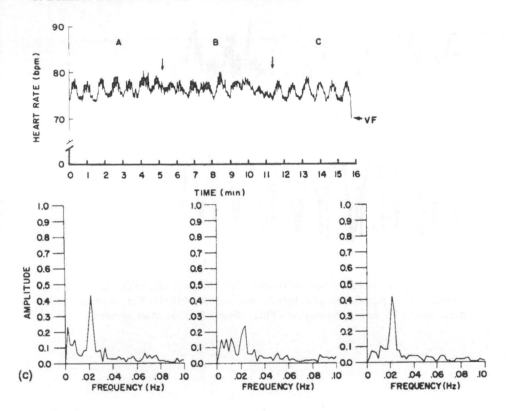

(c)

ately to a quick nearby solution (see Figure 9 for the 1-dimensional case). In the N-dimensional case, conventional algorithms usually settle for a local, but not a global, minimum. Surprising improvements came after the observation that computers cannot do all we might want if we demand perfection. Instead, if we allow computers to make mistakes—as humans do—we might be able to get them to do better (Kolata, 1984).

To implement these ideas, numerical analysts (Press et al., 1987) add a generator of random changes to the system configuration, so that the system will sometimes make the "mistake" of going uphill and subsequently move on to another minimum energy state.

This probabilistic algorithm is slower than conventional algorithms that look for a quick payoff; but there is an analogy in nature. For instance, at high temperatures the molecules of a liquid move freely with respect to one another. If the liquid is cooled slowly, thermal mobility is lost, and the atoms are often able to line themselves up and form a pure crystal that is completely ordered in all directions. (A crystal is the state of minimum energy for this system.) The amazing fact is that, for a slowly cooled system, nature is able to

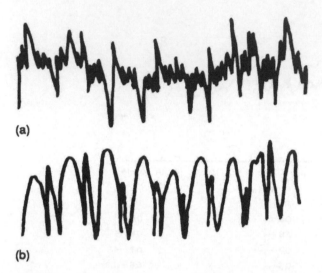

(a)

(b)

Figure 8 Brain waves of two individuals. (a) Normal adult EEG; (b) spike wave complex of an epileptic seizure. Original data are not available. The graphs are reproduced from a TV screen and may be a little different from the exact pictures.

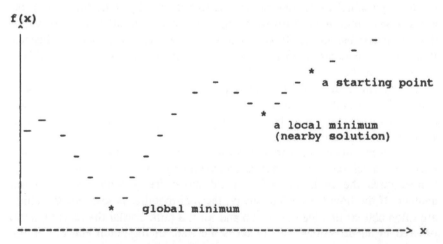

Figure 9 A search for the global minimum.

find this minimum energy state. On the other hand, if a liquid metal is cooled quickly or "quenched," it does not reach this state but rather ends up in a polycrystalline state having somewhat higher energy (Press et al., 1987).

The probabilistic law that governs the thermal equilibrium at temperature T is the so-called Boltzmann probability distribution:

$$\text{Prob}(E) \propto \exp(-E/kT),$$

where E is an energy state, and k is Boltzmann's constant. According to this probability law, sometimes nature does make the "mistake" of being in a high energy state. But in the process of slow cooling, there is a good chance for the system to get out of a local minimum in favor of finding the real minimum.

EXAMPLE 3 In certain engineering applications, a technique called "artificial dither" is often applied for the purpose of smoothing the nonlinearity in a feedback system, as shown in Figure 10. "Dither," according to the Webster's dictionary, means a confused and excited condition, a prohibited word to certain quantitative social scientists. But in engineering applications, a calculated dither can be profitably employed in the compensation of nonlinear control system, often for the purpose of linearizing the highly nonlinear force-velocity effects due to friction. Applications of dither have a very long history. Some examples are: coulomb friction, threshold, the controlling of small guided missiles by spoilers, position servo incorporating a photocell error detector, extremal control, adaptive control, signal stabilization, etc. (Gelb and Vander Velde, 1968).

Certain dither techniques use a *random*, nonlinear input that can be described as follows (Gelb and Vander Velde, 1968, p. 320):

$$x(t) = B + r(t) + A * \sin(wt + \theta),$$

where r(t) is a stationary Gaussian process; theta is a random variable having a uniform distribution over an interval of 2π radians; and B, A, and w are deterministic parameters. By careful calculation, this type of input may yield a linear approximator that produces a power density spectrum at its output equal to that of the nonlinearity it approximates. In other words, the intentional injection of random noise is indeed a fostering of restrained chaos in the sys-

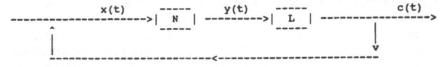

Figure 10 Feedback system in "artificial dither." x(t): input; c(t): control variable; N: nonlinear part; y(t): output; t: time; L: linear part.

tem. As a result, the random noise has the stabilizing effect of smoothing non-linearities and consequently suppressing certain undesired periodic behavior.

The morals of Examples 2 and 3 are that "mistakes" and tempering of a system are not always bad, and that perturbation by random inputs should not be outlawed altogether.

EXAMPLE 4   Rapp and co-workers (1986) examined the brainwave of a human resting [(Fig. 11(a)] to that of the same person who was instructed to count backwards from 300 in steps of 7 [Fig. 11(b)]. The two time series appear quite different: one is relatively smooth, while the other is chaotic. The power spectra of these two different series, however, are quite similar (Fig. 12). According to the ordinary practice (see examples in Chap. 1, Sec. VI), one might declare that the difference between the two series is *not* significant. Instead, Rapp and co-workers calculated the fractal dimensions (D's) of the series (Fig. 13), where C_n is the correlation integral (see details in Rapp et al., 1986). The fractal dimension associated with mental arithmetic is indeed higher than that associated with a human resting.[4] To draw a further contrast, Rapp and co-workers plotted the Poincare maps shown in Fig. 14, which show a remarkable difference between the two brain-waves. In recent studies, Rapp (February, 1990, quoted in *Omni*, pp. 43–48) found that the greater the mental challenge, the more chaotic the activity of the subject's brain. In other words, chaotic activity may be an asset in problem solving.

"You want to be able to scan as wide a range of solutions as possible and avoid locking on to a suboptimal solution early on," Rapp explained. "One way to do that is to have a certain amount of disorderliness, or turbulence, in your search." This strategy has indeed been implemented in numerical analysis, as we have seen in the example of "simulated annealing."

Another potential application of the above analysis is the design of electrode-studded headware for air traffic controllers, nuclear plant operators, radar monitors, and other professionals whose unwavering vigilance is critical to the safety of a large sector of the public (Rapp, 1990). In such cases, a loss of *healthy variability* in neural activity may be used as an indication of the loss of mental alertness. Alarms or electrical shocks may be needed to wake up the operators and avoid catastrophic accidents.[5]

In another line of research, Garfinkel examined tremors and EMGs (electromyograms) of victims of stroke and Parkinson's disease. Tremors, according to Garfinkel (1989, NOVA #1603), "may look random to the untutored eye, but in fact, if you look at it, they're in fact quite periodic." In subsequent studies, Garfinkel (February, 1990, quoted in *Omni*, pp. 43–48) concluded that patients with normal motor control had nerves that pulsed in a chaotic fashion, whereas the EMGs of spastic patients demonstrated much more regular bursts of electrical activity.

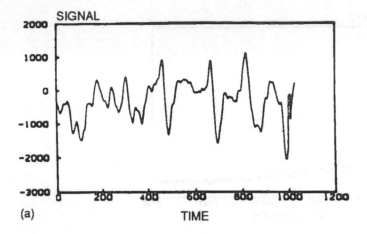

(a)

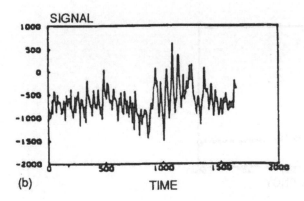

(b)

Figure 11 (a) Brain wave of a person at rest; (b) brain wave of same person counting backwards from 300 in steps of 7. Copyright © 1986 by Springer-Verlag. Reprinted by permission.

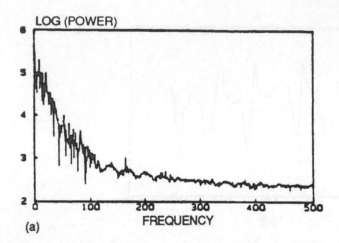

(a)

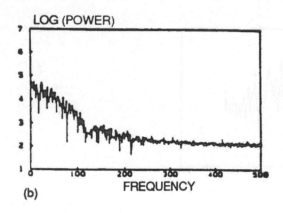

(b)

Figure 12   (a) Power spectra of a person at rest; (b) Power spectra of same person counting backwards from 300 in steps of 7. Copyright © 1986 by Springer-Verlag. Reprinted by permission.

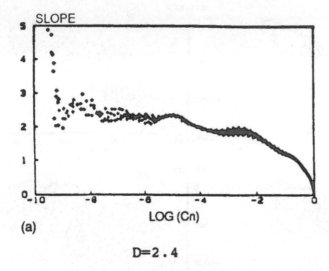

(a)

D=2.4

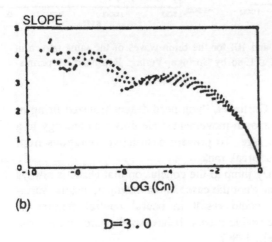

(b)

D=3.0

Figure 13 Slope vs. the logarithm of the correlation integral. Copyright © 1986 by Springer-Verlag. Reprinted by permission.

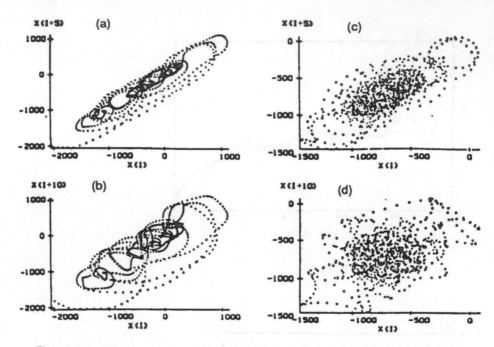

Figure 14   Delay maps (lag = 5 and 10) for the brain waves of the same person in two different situations. Copyright © 1986 by Springer-Verlag. Reprinted by permission.

"Contrary to intuition," says Garfinkel, "you need desynchronized firing of nerve cells in order to achieve smooth movement." He draws an analogy to a platoon of soldiers crossing a bridge. To prevent destructive vibrations from collapsing the bridge, the soldiers break rank.

At this point, some readers may jump to the conclusion that chaos is always beneficial in neural systems. That's not the case. As an example, chaotic variation in brain dopamine levels could result in neural control failures of Parkinson's disease and in the unstable mental behaviors that are characteristics of schizophrenia. (Rapp, et al., 1986).

To sum up, instead of charging into disorder, scientists are looking for constrained randomness, or a harmony between order and chaos. After all, the reason why science works is because there is order in nature.

NOTES

1. Even in the recruitment of graduate students, there are much concealed human judgments in the construction of the linear regression.
2. The ECG's one usually sees in a hospital exhibit a P-Q-R-S-T pattern. The time series we discuss here is obtained by measuring the length of consecutive R-peaks.

3. In certain engineering applications, (e.g., in the transmission of electical signals), noise cannot be brought into harmony with signals.

4. The standard errors of the D's are respectively .2 and .3, estimated by eyeball examination (Rapp et al., 1986). If we assume that the D's are normally distributed, then z = 1.67 and P = 5%, an indication that chance variation may be at work.

   Note that the statistical properties of such estimates, to my knowledge, have not been rigorously examined. Efforts along the line of Cutler and Dawson (1990, *The Annals of Probability*) may yield more satisfactory answers than those currently reported in scientific literature.

5. A similar device may also be used to prevent a driver from becoming a traffic statistic. As reported by the U. S. Department of Transportation, at least 40,000 traffic accidents a year may be sleep related. Further, more than 20% of all drivers have fallen asleep behind the wheel at least once (the *New York Times*, May 15, 1990).

REFERENCES

Bak, P. and Chen K. (1991). Self-Organized Criticality. *Scientific American*, Vol. 264, No.1, 46-53.

Cutler, C.D. and Dawson, D.A. (1990). Nearest-Neighbor Analysis of a Family of Fractal Distributions. *Annals of Probability*, Vol. 18, No. 1, 256-271.

Freedman, D.A. and Zeisel, H. (1988). Cancer Risk Assessment: From Mouse to Man. *Statistical Science*, Vol. 3, 3-56.

Gelb, A. and Vander Velde, W.E. (1968). *Multiple-Input Describing Functions and Nonlinear System Design*. McGraw Hill, New York.

Gleick, J. (1987). *CHAOS: Making a New Science*. Viking, New York.

Glymour, C., Scheines, R., Spirtes, P., and Kelly, K. (1987). *Discovering Causal Structure: Artificial Intelligence, Philosophy of Science, and Statistical Modeling*. Academic Press, New York.

Goldberger, A.L., Findley, L.J., Blackburn, M.R., and Mandell A.J. (1984). Nonlinear Dynamics in Heart Failure: Implications of Long-Wavelength Cardiopulmonary Oscillations. *American Heart Journal*, Vol. 107, No. 3, 612-615.

Goldberger, A.L., Bhargava, V., West, B.J., and Mandell A.J. (1985). Nonlinear Dynamics of the Heartbeat. *Physica*, 17D, 207-214. North-Holland, Amsterdam.

Goldberger, A.L., Bhargava, V., West, B.J., and Mandell A.J. (1986). Some Observations on the Question: Is Ventricular Fibrillation "Chaos"? *Physica*, 19D, 282-289. North-Holland, Amsterdam.

Goldberger, A.L. and West, B.J. (1987). Applications of Nonlinear Dynamics to Clinical Cardiology. *Annals of the New York Academy of Sciences*, Vol. 504, 195-212.

Goldberger, A.L., Rigney, D.R., Mietus, J. Antman, E.M., and Greenwald, S. (1988). Nonlinear Dynamics in Sudden Cardiac Death Syndrome: Heartrate Oscillations and Bifurcations. *Experientia* Vol. 44, 983-987. Birkhauser, Boston, Massachusetts.

Harris, M.(1980). *Cultural Materialism: The Struggle for a Science of Culture*. Vintage, New York.

Kolata, G. (1984). Order out of Chaos in Computers. *Science*, Vol. 223, 917-919.

Press, W.H., Flannery, B.P., Teukolsky, S.A., and Vetterling, W.T. (1987). *Numerical Recipes: The Art of Scientific Computing*. Cambridge University Press, Cambridge, Massachusetts.

Rapp, P.E., Zimmerman, I.D., Albano, A.M., deGuzman, G.C., Greenbaun, N.N., and Bashore, T.R. (1986). Experimental Studies of Chaotic Neural Behavior: Cellular Activity and Electroencephalographic Signals. *Nonlinear Oscillations in Biology and Chemistry*, edited by H.G. Othmer. Springer-Verlag, New York.

Voss, R.F. (1988). Fractals in Nature: From Characterization to Simulation. *The Science of Fractal Images*, edited by H.O. Peitgen and D. Saupe, Springer-Verlag, New York, pp. 21–70.

Chapter 8

The Riddle of the Ubiquitous Statistics

In 1980 the National Science Foundation (NSF) published a projection of job openings in science and engineering in the years 1978–1990. The projection estimated job openings for statisticians at 19,000, with a supply shortage of 11,000 over a 12-year period. As a comparison, the job openings for mathematicians were estimated at 3,000, with an oversupply of 126,000.[1]

This NSF publication has been cited in *Amstat News* (1981), *The American Statistician* (1983), *JASA* (1982, p. 5), and in a variety of promotional materials that are aimed at luring students into the statistical profession. The promotional efforts are in general accompanied by an optimistic assumption that statistics is useful in every part of our lives, and that it is beneficial for students to study statistics or enter the profession of statistics.[2] An example of this rosy description of statistics is the following (Anderson and Sclove, 1986):

> Statistics enters into *almost every phase of life* in some way. A daily news broadcast may start with a weather forecast and end with an analysis of the *stock market*. In a newspaper at hand we see in the first five pages stories on an increase in the *wholesale price index*, an increase in the number of habeas corpus petitions filed, new findings on mothers who smoke, the urgent need for timber laws, a state board plan for evaluation of teachers, popularity of new commuting buses, a desegregation plan, and sex bias. Each article reports some information, proposal, or conclusion based on the organization and analysis of numerical data. [emphasis supplied]

That is, statistics is everywhere. But a curious fact is that the discipline of statistics lacks visibility and influence, as pointed out in an ASA Presidential Address, "The Importance of Statistics" (Marquardt, 1987, *JASA*). Marquardt maintains that "the actual content of many jobs throughout society has more to do with statistics than with any other single subject matter field." In this job market, Marquardt observed, there are many competitors (as opposed to statisticians):

> engineers (industrial, electrical, mechanical, electronic, chemical, etc.)
> physicists
> business professionals, such as those with M.B.A. degrees
> operations researchers
> mathematicians
> social scientists
> computer specialists

One may wonder why these people are classified as "competitors" to statisticians. According to Marquardt,

> They do such things as deciding what data are needed for a specific problem and how to collect the data. They summarize the data in tabular and graphical form. They analyze the data. They develop predictive models from the data. They derive conclusions and recommend courses of action, all based largely on the data.

"What are they doing on the job?" Marquardt asked. "They are doing statistics for a large proportion of their time." Marquardt complained, "if statistical tools and functions are the primary components of many jobs, why can't people trained in statistics hold some of those jobs?"

Marquardt's view of statistics is similar to that of Hahn and Boardman, two leading figures in statistical quality control. According to Hahn and Boardman (1987), making decisions needs data, and data means statistical methods.[3]

To concerned statisticians, the promotional efforts of Marquardt, Hahn, etc. are well-intended, highly admirable, and should be put forward with even greater force. Nevertheless, the facts that statistics is everywhere and that statistics as a discipline lacks visibility and influence are hard to explain. After years of confusion I finally came to realize that the noun "data" in statistics is not equivalent to the noun "data" in other contexts. For example, in computer science, "data base," "data structure," and "data-base management" all contain the noun "data," but in a sense that is foreign to most statisticians. For another example, a telephone book contains a mass of data that are useful for many decisions, but statistical inference plays no role in it.

In Section I of this chapter, we will take pains to re-examine certain misconceptions about the ubiquitous statistics. Nevertheless, an *optimistic* prediction for the future of statistics will be provided at the end of that Section.

I. INFORMATION AND MISINFORMATION

First there is a great divide between measurements and statistics. When we examine the whole spectrum of statistical applications, on one end we find that the measurements in social, behavioral, or economic sciences are very rough, and the researchers call all numerical values "statistics," as if the name itself will enhance the accuracy of a poor measurement. On the other end of the spectrum, physicists, engineers, and chemists, who keep improving their instruments, call all experimental readings "data," but they seldom need to perform a complicated statistical test.

To further illustrate the situation, consider the wisdom of a famous statistician, C. R. Rao (from DeGroot, 1987):

> I don't think we are very successful with statistical methods in psychology or even in economics. Possibly what is wrong with the economists is that they are *not* trying to measure new variables which cause economic changes. That is far more important than dabbling with whatever data are available and trying to make predictions based on them. [emphasis supplied]

Rao's message is that one has to pour more concrete into the foundations of one's statistical models. The concrete is subject-matter knowledge and quality data.

Note that the two most important (and reliable) methodologies in statistics are randomized experiments and sample surveys. Also, randomization has to do with the quality of data (i.e., quality of measurement), not with the textbook statistical formulas. Outside the textbooks, most data are *not* obtained by randomized experiments or sample surveys. Given this kind of data, a person who is rich in subject-matter knowledge (or street smarts) is likely to see more than a mathematical statistician who is thrown into the situation. For this reason we contend (as opposed to Marquardt) that many business or industrial jobs should be held by people in these fields (not by statisticians), even if a job requires that a large portion of time be spent doing data reduction and graphic representation.

The following examples are chosen to illustrate further the relationship between statistics and the real world. The examples may help one to understand why statistics is everywhere but statisticians are not.

EXAMPLE 1 If one is fond of data, one should take up accounting. In this area the data are very rich and important: They show financial records such as what cash and equipment an organization has, what debts it owes, etc. This type of data provides information to executives and government officials so that they can make sound decisions. These decisions, with no exaggeration, "affect the welfare of all citizens" (*Statistical Science*, 1989). But formal statistical inference is completely useless in this context.

In recent years statistical methods have been brought to bear in an attempt to help the analysis of audit data (which is concerned with examing and reviewing accounting information to determine whether it is reasonably accurate and presented in an understandable fashion). According to a report by the Panel on Non-Standard Mixtures of Distributions (Tamura, 1989), the existing models for auditing data include the three-parameter gamma distribution, the geometric distribution, parametric Bayesian models, non-parametric Bayesian[4] models, etc.

Despite such heavy machinery, the real progress is dismal. The report also points out that one-sided confidence intervals are more appropriate in the analysis of audit data than two-sided confidence intervals. This conclusion, in my opinion, further casts doubt on the accuracy and value of the estimation.

In summary, accounting data defy easy analysis along the lines of standard (or nonstandard) statistical inference. At this writing the statistical community still cannot claim the vast territory of financial records and auditing data.

EXAMPLE 2   Statistics is often perceived as being capable of providing useful tools for sound decision-making. Innumerable books have been written, and thus yield an impression that there *must* be something in "statistical decision theory" that one can use in real-life applications.

The kernel of truth is that statistical decision theory does not deal with ordinary decisions. Rather, it deals mainly with the properties of Bayesian estimators for the population average under certain loss functions.

Suppose that you are going to buy a car or a house—certainly a big decision that you don't want to regret soon afterward. Try any book on Bayesian decision theory; it is guaranteed that none of the books will help you in buying a car or a house.

In short, "statistical decision theory" is totally different from "ordinary decision theory." Statistical decision theory is a body of esoteric knowledge that is of interest mainly to academicians. Just like certain intricate equations in modern physics, statistical decision theory in its current state belongs in an ivory tower that has little rooting in the rest of the world.

EXAMPLE 3   In recent years, the ivory-tower "statistical decision theory" has transcended its traditional concern, linking up with research effort in artificial intelligence (AI) and expert systems. As much of the world knows by now, the goal of AI is to aid decision makers in all areas of human endeavor. Even if only part of the goal of AI materializes in specific areas (such as medicine, public policy, business, military affairs), the payoff of the ivory-tower activities would rival the discoveries of Dirac or Einstein.

On the surface decision-making involves data and uncertainty; thus statistics and probability theory ought to play a vital part in the field of artificial intelligence. But as pointed out by the editor of *Statistical Science* (February, 1987), "such has not proved to be the case." For instance, if one looks through the three-volume *Handbook of Artificial Intelligence* (Barr and Feigenbaum, 1981, 1982; Cohen and Feigenbaum, 1982), one can barely find such things as statistical learning, probability reasoning, uncertainty in representation of knowledge, and the like.

In the late 70s and early 80s, the emerging field of artificial intelligence stood out like a king. However, in private, some computer scientists have commented that the emperor really has no clothes. These comments were not made out of jealousy, but rather were rooted in a deep understanding that decision-making is a highly conditioned activity, prey to the effects of many factors that cannot be controlled by any mechanical procedures and cannot be predicted by any mathematical formula.

Instead of the oversweeping goal of AI in modeling everyday reasoning, expert systems that are designed for specific fields are more realistic. Developments have proved to be successful in certain medical diagnoses. But as Spiegelhalter (1987) wrote, "the development of expert systems in medicine has generally been accompanied by a rejection of formal probabilistic methods for handling uncertainty."

New developments that use Bayesian methodology or belief function are underway. Some general obstacles have been pointed out by Shafer (1987):

Probability judgment in expert systems is very much like probability judgment everywhere else.

Almost always, probability judgment involves not only individual numerical judgments but also judgments about how those can be put together.

[If progress is ever made in using probability in expert systems], these systems will be *intensely interactive*. They will depend on the human user to design the probability argument for the particular evidence at hand. [emphasis supplied]

Shafer maintains that it is neccessary to implement heuristic probability judgments assigned by human users, despite the fact that such heuristics often lead to systematic mistakes or biases. In defense of this position, he observes that when we face up to the AI problems, "we see that the heuristics are really all we have."

Shafer's honest assessment of probability judgments is a clear indication that the academic version of "decision-making" is far from the real-world version. This is good news for researchers in the field, but it is bad news for those who believe that decision-making needs data, and data is equivalent to statistical methods.

• • • • • • • • • • • •

Despite the discussion in the previous examples, it is not our stance that statistics is utterly useless. On the contrary, we have tried to include *constructive uses* of statistics in such examples as

a modern version of Fisher's puzzle (Chap. 2, Sec. V),

graphic presentation and time-series analysis of Old Faithful data (Chap. 4, Sec. V)

Shewhart control chart (Chap. 3, Sec. IV)

parameter design in quality control (Chap. 8, Sec. II)

descriptive statistics (Chap. 5, Sec. I.F.)

Bayesian analysis (Chap. 6, Secs. III and IV)

chi-square test (Chap. 2, Sec. V)

non-linear modeling (Chap. 5, Sec. II.E)

spectral estimation (Chap. 4, Sec. V and Chap. 7, Secs. I and II) and

quantum probability (Chap. 6, Sec. II).

These examples are intended to illustrate that *statistics as a discipline is infinitely rich in its subtlety and esoteric beauty.*

For the readers who want to know more about authentic applications of statistics in science, a good source is *Statistics* by Freedman, Pisani, and Purves (1978). For those who want to learn about ingenious applications of statistics in industry and quality control, an excellent book is *Process Quality Control: Troubleshooting and Interpretation of Data* by Ott (1976).

Good examples of statistics in everyday life are also plentiful. For instance, according Deming (1978), in Japan the whole month of October is a season of statistical observance. Here are some specific examples:

Children in their lessons in October describe their activities for the day in terms of numbers: passed by, on the way to school, so many automobiles of a certain type; walked to school in an observed number of minutes; her mother spent so many hours with her last night, of which a certain proportion were spent concerning arithmetic, another proportion on geography, another proportion on hiragana (Chinese ideograms).

For another example, Deming (1978) mentioned that in Japan a national prize was given to five *7-year-olds* for a graph creation titled "Mom, play with us more often." The work was the result of a survey of 32 classmates about the frequency with which mothers play with their offspring and the reasons given for not doing so. The most often heard excuse: "I'm just too busy."

In addition to these beautiful examples, I would like to share an anecdote from my own experience. When I started my teaching career, I used to feel heart-broken during the grading of midterm or final exams. In the middle of the grading, I railed against the poor performance of the students. "What the Hell? This problem is merely a simple application of the Central Limit

Theorem. Why can't they do it?" Such complaints ran throughout the whole session of grading. But a few minutes later, I graphed the test scores in a stem-leaf plot:

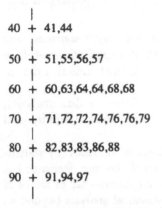

```
    |
40 +  41,44
    |
50 +  51,55,56,57
    |
60 +  60,63,64,64,68,68
    |
70 +  71,72,72,74,76,76,79
    |
80 +  82,83,83,86,88
    |
90 +  91,94,97
    |
```

As the graph revealed, the overall performance of this class was as normal as that of other classes taught by myself or by other professors. There was simply no need to feel bitter at all.

This is a case where the psychological involvement of a teacher overwhelmed his judgment. And a statistical artifact was the best rescue. Similar examples can be found in every walk of life. For example, when we have bitter feeling about our family members or our work environment, a frequency distribution of related events may easily diffuse a potential conflict.

At this juncture, one may argue that statistics are indeed ubiquitous. Our response to this conclusion is as follows. First, statistics are in fact everywhere, but valid statistical inference is not. As we have taken pains to explain, statistical inference applies only to well-defined chance models. In everyday life, such models are seldom appropriate.

Second, everyday statistics (or elementary "descriptive statistics") rely on a person's subject-matter knowledge. The hard part of getting such statistics depends on the skill and instinct of the person collecting the data. Putting the data in a statistical graph is easy; it requires only the brain power of a 7-year-old.

·   ·   ·   ·   ·   ·   ·   ·   ·

In the Preface we mentioned that there are about 40,000 scientific journals currently being published regularly (*The New York Times*, June 13, 1988). Assume that 20% of journal articles involve data collection and statistical analysis. Also assume that on the average one out of three submitted papers is accepted for publication (this ratio is substantially lower than one-third in prestigious journals). Putting these numbers together, we believe that the scientific community would need an army of competent statisticians to ensure the validity of statistical analysis in research publications.

In addition to this service, statisticians possess a wealth of techniques for exploratory data analysis (EDA) which includes Tukey's classic EDA (1977), a new generation of techniques for exploring data tables, trends, and shapes (Hoaglin, Mosteller, and Tukey, 1985), advanced descriptive statistics, and quasi-inferential statistics discussed in Chapter 2. Skillful applications of these techniques may help to uncover important patterns and causal links in scientific investigations or industrial applications.

With the techniques in both EDA and CDA (confirmatory data analysis), we don't understand why the discipline of statistics lacks visibility and influence as lamented by Marquardt (1987, *JASA*).

Perhaps the reason is that the teaching and practice of statistics (by statisticians!) are both too much and too little. It is too much, because statistics has been equated with "data" and has been touted as ubiquitous—but in fact it is neither. This tendency stretches the limitation of statistical analysis beyond its possibility.

It is too little, because statisticians themselves do not pay enough attention to the stochastic assumptions in their data analyses (see examples in Chapters 2–4). As a consequence, scientists from all disciplines (in academics and in industry) learned this attitude and are using over-the-counter statistics in every corner of their research activity without feeling the need of a statistician.

In conclusion, we believe that statistics is a scientific discipline that should (and will) gain influence and visibility, but statisticians first have to divorce themselves from the current practice of statistics in soft sciences. The practice is self-delusive and indeed is a disservice to the discipline of statistics. This divorce, in our opinion, is the first step necessary to restore statistics as a scientific discipline.

II.   STATISTICAL QUALITY CONTROL AND THE CONFLICTING TEACHINGS OF Q. C. GURUS

Something alarming has happened to the United States of America since the 1970s: Her worldwide market has dropped by 50% in the areas of automobiles, medical equipment, computer chips, and consumer goods such as color TVs, cameras, stereos, microwave ovens, etc.

A portion of this drop is usually attributed to the Japanese for their aggressive market strategy and the superior quality of their products. But other factors may contribute as well. For instance, the United States ranked number one in her productivity as compared to other industrial countries in 1950 (here the productivity is defined as the average annual growth in manufacturing output per hour):

Rank	Country
1	**United States
2	Belgium
3	West Germany
4	France
5	Norway
6	Canada
7	Sweden
8	Italy
9	United Kingdom
10	Denmark
11	Japan

If you ask an average American (or American student) to rank the productivity of the USA in 1987, the answer is likely to be 2 or 3, right after Japan and/or Germany. But the correct answer is more disappointing than that:

Rank	Country
1	Japan
2	West Germany
3	Italy
4	France
5	Norway
6	Belgium
7	Denmark
8	Sweden
9	United Kingdom
10	Canada
11	**United States

The fact that the productivity and the quality of American goods are trailing behind other industrial nations worries the political and business leaders of this country. Some initiatives were taken to raise the awareness of the problem and to implement potential solutions (such as using statistical methods in the battle against poor quality). In one instance the famed statistician, Dr. Edwards Deming, was awarded a National Medal from the President of the United States "for his forceful promotion of statistical methodology, for his contribution to sampling theory and for his advocacy to corporations" (*The New York Times*, June 26, 1987).

Deming is a statistician whose early books such as *Sampling Theory* (1953, Wiley) and *Statistical Adjustment of Data* (1938, 1943, 1964) are recognized as classic works. In 1950 Deming was recruited by General McArthur in the effort to rebuild Japan from the rubble of the war. In Japan Deming taught elementary statistical techniques and Shewhart control charts to Japanese engineers and managers at all levels. Deming's lectures were received with overwhelming enthusiasm. A national Statistics Day was established by the Japanese, and has been observed by Japanese yearly ever since.

As much of the world knows by now, the Japanese have performed a miracle by keeping the quality of their consumer products at a superior level. According to Deming (1982, pp. 99–102), this miracle was due to the following four factors: (1) Japan's statisticians, (2) the Union of Japanese Science and Engineering, (3) the teaching of techniques, and (4) conferences with top management. (This analysis is different from mine. See footnote 5.)

Deming enjoys (and deserves) the great honor he received when the Japanese established the Deming prizes for high achievement in quality improvement. His teachings had long been ignored by his countrymen from the 1950s to the early 1970s. Later, with the enormous success of Japanese quality associated with his name, his return to his homeland was cheered by the zealots full of unrealistic expectations about the role of statistical methods and statisticians in quality control.

It is sad to note that these unrealistic expectations were initiated by Deming and his fanatic followers such as Joiner (1985). For instance, Deming contended that a strong "statistical leadership" is essential in any company or organization, and recommended that "there will be in each division a statistician whose job (it) is to find problems in that division, and to work on them" (Deming, 1982, p. 354). With regard to the focus of this *statistical leadership*, Deming (p. 355) suggested the following schematic plan for "statistical organization" indicated in Fig. 1. Deming wrote, "it is clear that no plan will work . . . without competence and confidence in the statistical leadership, and without people in the divisions that have a burning desire to improve their statistical work." Furthermore, "The advantages of the plan recommended here cannot be questioned. It works."

Deming's writings on quality control, if judged fairly, contain much useful information and numerous sparks of wisdom. For instance, he considered the difference between theoretical statisticians and *practical statisticians*, and concludes that the latter are the real hazard, citing Thomas Huxley's famous quote, "The practical man practices the errors of his forefathers."

Nevertheless, his preaching of so-called "statistical leadership" does not seem to have won the hearts of certain statisticians. Pointing to the worst

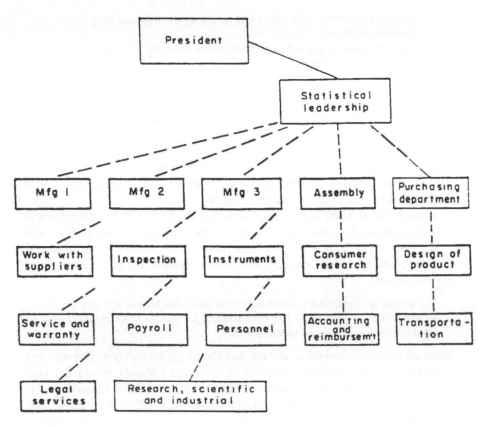

Figure 1   Schematic plan for statistical organization. Source: W. E. Deming (1982). *Quality, Productivity, and Competitive Position.*

scenario, Godfrey (1985) wrote, "I shudder at the thought of my statistics colleagues running into executive offices and telling the managers how to run the company." Godfrey maintains that

> We as statisticians should focus our efforts on what we know best. We do have much to give ... [but] we must stick to our areas of expertise.

This conclusion and the related examples in Godfrey's article are the real food for thought to applied statisticians (or would-be applied statisticians) who are excited or confused by the so-called "statistical leadership." Note that "our areas of expertise" include many technical advances such as:

Robust parameter design
Improved Taguchi's on-line quality control
More efficient method for estimating SNR (signal-to-noise ratio)

Factorial experiments with nearly orthogonal arrays and small runs
Nonlinear estimation
Regression analysis applied to artificial neural networks
Wavelet transform and its applications in statistics

It should be pointed out that "statistical control charts," strictly speaking, are *not* our areas of expertise. The reason can be highlighted by H. F. Dodge's statement that "statistical quality control is 90% engineering and only 10% statistics" (Grant and Leavenworth, 1980, p. 88).[6] In other words, an effective use of the statistical control charts requires one's knowledge of physics, engineering, and the subject matter.[7] As a result, a statistician's areas of expertise are most useful in the Department of Research and Development, not in the production lines or in the executive offices.

I would like to make one more observation on the role of the statistician in decision-making. It is interesting to review the position Deming took on this issue back in the 1960s. In a special address to the Institute of Mathematical Statistics, Deming (1965) strongly suggested a nonadvocacy role for a consulting statistician:

> The statistician should not recommend to the client that he take any specific administrative action or policy.... The statistician, if he were to make recommendations for decision, would cease to be a statistician.

Some statisticians comment in private that Deming is now only a salesman, not a statistician. This comment may not be fair, and certainly it will not (and should not) affect Deming's place in history. But Deming's followers, who exaggerate the role of statistics, should think twice before they blow their horns again.

•   •   •   •   •   •   •   •   •

Most experts on quality control do appreciate statistical methods as useful tools in this field. But very few of them exhibit the same zeal for statistics as do Deming and his followers. In this section we will discuss some of the conflicting teachings of the QC experts. Among the experts who will be quoted here are (1) Juran, (2) Crosby, (3) Ishikawa, and (4) Deming. This discussion is not meant to disparage these QC gurus; rather, it is a comparative study that promotes *a skeptical eye and an appreciative mind* toward different schools of thought. The moral is that one has to be careful in subscribing to any specific philosophy of management quality control.

We begin with Dr. Joseph Juran, the person who coined the phrase, "Pareto principle," to capture a universal phenomenon—vital few, trivial many—in quality control (and in many other areas too). He is also the first person to have asked his audience (top managers, engineers) this beautiful question: "Does better quality cost more or less?" From most people the response would be an irrefutable "More!" Juran's genius was to point out that there are two different notions of quality: quality of design and quality of conformance. If one is talking about quality of design, then it is certain that better quality costs more. But if one is talking about quality of conformance, then better quality will cost less (Juran, 1979).

Juran made his first visit to Japan in 1954, at the request of the Union of Japanese Science and Engineering. His influence on quality control can be seen in his massive *Quality Control Handbook* (1951/1979; three inches thick). In one place Juran uses a mathematical model to illustrate a certain aspect of the cost of quality (see Fig. 2). This model makes perfect sense to any "reasonable" mind. But one becomes puzzled reading Philip Crosby's book.

Crosby is the president of the Crosby's Quality College in Winter Park, Florida (Crosby, 1984). The College has attracted more than 15,000 executives since it began in 1979. Prior to 1979 Crosby was corporate vice-president of ITT (International Telephone and Telegraph Company). He has had 34 years of first-hand experience in quality improvement, and was responsible (while at ITT) for worldwide quality operations for 14 years. Unlike Deming, Juran, or Ishikawa, who have strong academic backgrounds, Crosby worked his way up from line inspector. His writings on quality control are the most colorful and eye-opening among those of all QC gurus.

According to Crosby it is a mistake to talk about the economics of quality. To him the cost of quality is the expense of nonconformance—the cost of doing things wrong. He emphasizes that quality means quality of conformance.

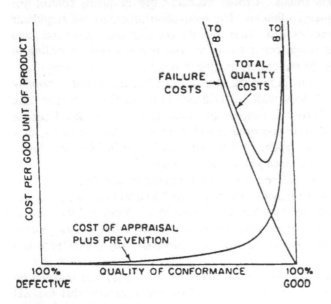

Figure 2 Model for optimum quality costs. Copyright © 1979 by McGraw-Hill Book Company. Reprinted by permission.

No more, no less. Quality is not elegance, not quality of design. Quality means that you set up specifications, and do it right at the beginning. Finally, the standard of performance is "zero defects." Nothing else is acceptable. His insistence on "zero defects" has won him the nickname "quality fascist."

Crosby has a very rich array of arguments and examples to support his zero-defect philosophy. For instance, he maintains that "companies try all kinds of ways to help their people *not* meet the requirements" (Crosby, 1984, p. 75; emphasis supplied). Here is how they do it:

Acceptable quality level (AQL, 1%, 2.5%, etc.): This is established for suppliers as the acceptance plan for inspection or test people.

Shipped-product quality level (SPQL): This means a certain number of errors are planned. Refrigerators have perhaps three or four, computers eight or more, TV sets three or more.

On another front, it is common sense that motivation of the employees is important in quality improvement (Juran, 1979, Chapter 18). The motivation programs are now a multimillion-dollar business. The techniques include raising workers' self-esteem and making them more independent and assertive. In sharp contrast, Crosby emphasizes "demotivation" (1984, pp. 14–35). He asks: "Why do we need a special program to motivate our people? Didn't we hire motivated employees?" He continues, "If you think about it, they were well motivated when they came to work. That first day, when they reported in, there were nothing but smiles." Crosby maintains that in quality control one has to look for long-term solutions. The motivation programs are stimulants that get your employees excited. After a while, the employees are numb, and no real improvement would occur anymore. The real solution, according to Crosby, is inculcating the philosophy and the techniques of "zero defects."

Critics of Crosby maintain that the zero-defect (ZD) movement is nothing but chasing a ghost. Perhaps with this criticism in mind, Crosby wrote: "the only people who utilized the concept for what it was were the Japanese Management Association. In cooperation with Nippon Electric Company, they established the concept in Japan. (When I went to Tokyo in 1980 they had a party for me to celebrate sixteen years of ZD in Japan.)"

We now compare the different views of Ishikawa on this issue. Dr. Ishikawa is the inventor of the famous "fishbone chart" in quality control, and was a Professor of Engineering at the Science University of Tokyo and the University of Tokyo. He is now almost the spokesman for Japanese quality control. His book, *What is Total Quality Control? The Japanese Way* (1981/1985) is a best seller in Japan.

Ishikawa (1985, p. 151) observes that the ZD movement was initiated by the U. S. Department of Defense in the 1960s. He maintains that the ZD movement could not succeed, and that the QC community should not repeat the same mistakes. He wrote, "in some quarters, the United States has been strongly influenced by the so-called Taylor method. Engineers create work standards and specifications. Workers merely follow. The trouble with this approach is that the workers are regarded as machines. Their humanity is ignored." He maintains that

> Standards and regulations are always inadequate. Even if they are strictly followed, defects and flaws will appear. It is *experience and skill* that make up for inadequacies in standards and regulations. (p. 66) [emphasis supplied]

Ishikawa emphasizes that a respect for humanity is needed as part of management philosophy:

> Employees must be able to feel comfortable and happy with the company, and be able to make use of his capabilities and realize his potential.

This may sound like Utopia. But Ishikawa is confident that there are at least two ways to enhance humanity in a manufacturing environment: (1) recognition of and awards for achievements, and (2) the implementation of QC circles.

A QC circle, to put it in a capsule, is a small group of people who perform quality control activities within the same workshop. The idea of such group activity is hardly new. As a matter of fact, workers in other countries have similar arrangements to solve problems of defects or poor quality. But this simple idea, pushed by Ishikawa in 1962, evolves into a unique Japanese phenomenon that no one expected. For instance, projects completed by QC circles during its first decade (1962-1972) accumulated to about five million projects, with money saved at about $5,000 per project. The impact of the QC circle movement on Japanese quality has been simply phenomenal.

In the early 1970s the concepts of QC circle were brought to the United States and began to blossom soon after. In 1977, the International Association of Quality Circles (IAQC) was formed in the United States. Membership grew steadily in six years from 200 to 6,000. Conference attendance also grew, in the same time period, from 150 to 2700.

Ishikawa places great confidence in QC circles. He wrote (1985): "Where there are no QC circle activities, there can be no Total Quality Control activities." However, Deming observed that the implementation of QC circles is mostly a management fad in the United States. Each year hundreds of businesses in this country experiment with QC circles. Most of them fail immediately. A few reported success at an early stage, but burned out soon after. This is "a rediscovery of the Hawthorne effect," as well put by Deming (1982, p. 109). In fact, Deming maintains that "QC circles are a raging disease in America, and even worse in some other countries."

In Japan, QC circles are voluntary activities that are rooted in their tradition of teamwork and striving for perfection. In America, according to Deming, the same tool is used by managers as a quick cure-all, and is viewed by workers as a managerial gimmick that merely attempts to get them to work harder.

Deming, on the other hand, emphasizes a top-to-bottom reform. This philosophy goes against a common perception that poor quality is mainly the result of a poor work force. For instance, when it comes to the responsibility for poor quality, Deming believes that 85% of problems are related to the system, not the workers. This philosophy believes that an overall change in corporate culture and managerial approach is needed to regain the competitive edge of any quality-troubled organization. To implement this philosophy, Deming proposed his famous 14 points as a guideline for creating a new managerial climate.

This philosophy, however, is quite different from that of Crosby, who also proposed his own 14 points for reshaping corporate culture and attitudes. In some cases (see e.g., *Inc. Magazine*, May, 1985), we believe that it is Crosby's (rather than Deming's) philosophy that turned around a quality-troubled company. Nevertheless, we have no evidence that this will in general be the case.

•   •   •   •   •   •   •   •   •   •   •

In recent years, a Japanese engineer, Genichi Taguchi, proposed a new philosophy for quality problem solving, which emphasizes mathematical constructs of loss function and the design of experiments.

In this subsection we will discuss certain aspects of the Taguchi techniques and philosophy. One of these techniques is called parameter design, which includes many beautiful applications that are fascinating in their own right. To illustrate the basic principles of parameter design, we will discuss the following investigation conducted by Dr. Hilario Oh and Dr. K. Y. Im (1991) at General Motors Corporation, Warren, Michigan.

The study focused on the "snap-fit" fasteners that are used abundantly in cars. Of central concern here is the excessive variability in retentions (R) of a special type of fasteners:

2.29, 3.37, 3.86, 5.15, 5.24, 5.33, 5.77, 5.77, 7.18, 7.33, 8.10, 8.29, 8.29, 8.61, 9.44, 9.57, 10.73 (n = 17, mean = 6.72, SD = 2.37).

To reduce the variability in R, traditionally an engineer would tighten the tolerance of the diameter of the fastener (W), the diameter of the mating hole (D), and other related parts. But in this case, that option is not available because the variabilities of these parts have already been minimized to the levels that current technology would allow.

Instead of surrender, Drs. Oh and Im attacked the problem from a different angle. To begin with, they noticed that

$$R = K * J, \tag{1}$$

where K = the stiffness of the fastener and J = (W − D)/2 = the interference between the fastener and its mating part.

The nonlinear nature of R (as a function of K and J) provides the possibility for parameter design. Note that if K is fixed, then R is a linear function of J (Fig. 3). The shaded area on the R-axis represents the tolerance region of R. Theoretically, if one picks the middle of the tolerance region, then one could identify the J-value as follows (Fig. 4). However, in mass production, there are variabilities in both J and K (Fig. 5). Therefore the resulting values of R will be all over the place. To achieve a tighter control of R, there are two different approaches. The first approach is to manipulate the parameters K and J. In graphic form, this approach is equivalent to moving the wedge and the base J-value in Fig. 5 (Oh and Im, 1991).

The second approach is to conduct a mathematical analysis as this author has done below: Let S1 and S2 be the standard deviations of J and K, respec-

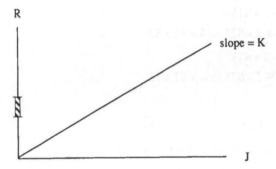

Figure 3 The linear relationship between R and J for the nonlinear function R = K * J.

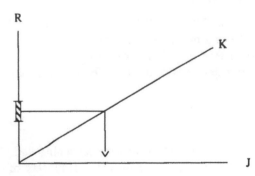

Figure 4 Identification of the J value from the middle of the tolerance region.

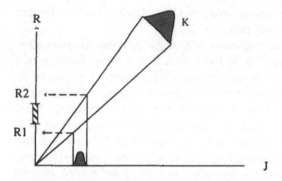

Figure 5   Graphic representation of manipulating parameters K and J.

tively, and R1 and R2 be the R-values indicated by the horizontal lines in the previous diagram. By formula (1), we have

$$R1 = (K - 2*S2)*(J - 2*S1)$$
$$= K*J - 2*(K*S1 + J*S2) + 4*S1*S2$$
$$R2 = (K + 2*S2)*(J + 2*S1)$$
$$= K*J + 2*(K*S1 + J*S2) + 4*S1*S2$$

Hence,

$$R2 - R1 = 4*(K*S1 + J*S2).$$

Let m be the target value of R, and note that $K*J = R = m$.
Therefore, we have

$$R2 - R1 = 4*(m*S1/J + J*S2), \tag{2}$$

By equation (2), one can minimize $(R2 - R1)$ by choosing

$$J = \sqrt{m*S1/S2} \tag{3}$$

In this case, m = 6, S1 = .06 and S2 = 5.09. Hence J = .27 and $(R2 - R1) = 10.83$, smaller than the original R2 − R1, which is 12.59. This constitutes the first phase of the parameter design in this case study.

   The second phase of this parameter design (Oh and Im, 1991) rests on the following mathematical formula:

$$K = (E/6)*\frac{w*t^3}{b*L^2} \tag{4}$$

where E is a constant (related to a special modulus), and the remaining variables are nominal dimensions depicted in Fig. 6. Raw data of 17 sets of fasteners are shown in the following table.

No.	Dimensions of Retainer and Slot (mm)						Estimated Values		
	t	b	w	L	D	W	J	K(N/mm)	P-Force(N)
1	0.33	0.42	3.10	4.38	5.98	6.70	0.36	19.38	8.29
2	0.33	0.42	3.35	4.29	6.00	7.00	0.51	20.14	8.61
3	0.25	0.42	3.20	4.17	6.01	7.00	0.51	9.02	3.86
4	0.33	0.42	3.40	4.38	5.98	6.90	0.46	19.38	8.29
5	0.33	0.38	3.20	4.29	5.98	6.60	0.31	22.38	9.57
6	0.33	0.50	3.20	4.29	5.98	6.70	0.36	16.79	7.18
7	0.25	0.46	3.10	4.25	5.98	6.70	0.36	7.88	3.37
8	0.21	0.38	3.30	4.33	5.98	6.85	0.43	5.36	2.29
9	0.29	0.46	3.20	4.33	5.98	6.80	0.41	12.03	5.15
10	0.38	0.46	3.20	4.38	6.00	6.75	0.38	25.09	10.73
11	0.30	0.38	3.15	4.20	5.98	6.90	0.46	17.04	7.33
12	0.30	0.49	3.20	4.32	6.00	6.95	0.48	12.39	5.33
13	0.36	0.52	3.10	4.13	5.98	7.00	0.51	21.96	9.44
14	0.30	0.45	3.25	4.32	5.98	6.95	0.48	13.42	5.77
15	0.34	0.49	3.15	4.18	6.00	6.95	0.48	18.84	8.10
16	0.30	0.45	3.15	4.32	6.00	6.75	0.38	13.42	5.77
17	0.30	0.50	3.20	4.30	6.00	6.85	0.43	12.19	5.24
Avg	0.31	0.45	3.20	4.28	5.99	6.84	0.43	15.69	6.72
Std	0.04	0.04	0.08	0.07	0.01	0.12	0.06	5.54	2.37

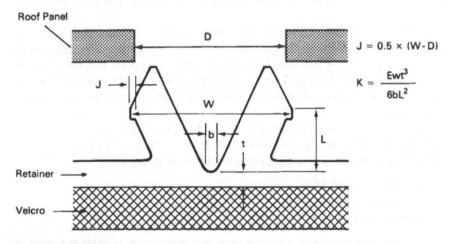

$$J = 0.5 \times (W - D)$$

$$K = \frac{Ewt^3}{6bL^2}$$

Figure 6 Graphical display of the "snap-fit" fastener. (Reprinted from Oh and Im, *Robust Design, A Need and an Opportunity*, 1991, presented in the 218th Meeting of the Institute of Mathematical Statistics. Courtesy of Oh and Im.)

A hierarchy of the attribution of the variabilities in R, K, and J is shown in Fig. 7.

In the upper-left corner, the graph indicates that every millimeter error in thickness t will result in 170 (N/mm) error in stiffness. Since the process for producing thickness is already at its best, further tightening of the stiffness variation has to be achieved through formula (4). Note that there are four independent variables in the formula, tangled in a highly nonlinear fashion. An analytic approach like formulas (2) and (3) would involve the 16 K-variables

$$K(i) = (E/6) * (w + .16) * (t \pm .08)^3 * (b \pm .08)^{-1} * (L \pm .14)^{-2};$$

$$i = 1, 2, \ldots, 16; \text{ plus signs throughout in } K(1);$$

and the search for the minimal values of

$$K(1) - K(i); i = 2, 3, \ldots, 16.$$

Such a task is strenuous (if not forbidden). A numerical approach to find the global minimum to desired accuracy is possible (see, e.g., Breiman and Cutler, 1989) but will be reported elsewhere.

A different approach is to examine the functional form of K(t,b,w,L) when some of the independent variables are fixed. For instance, if $t = .31$ (its mean value), $b = .45$, and $w = 3.2$, then K has an inverse-square relationship with L (see Fig. 8). From the graph, apparently a larger value of L will result in smaller variability in K.

In any event, by trial-and-error and smart guessing, Oh and Im chose a combination of nominal values of (t,b,w,L) and achieved the reduction of variations in K and R (Fig. 9). This constitutes the second phase of parameter design.

In the third phase, this author repeatedly applied formulas (1)–(3) as follows. Note that the original target value of R is 6 N-units, which is designed to prevent R from being less than 2 N-units (on the assumption that the SD of R is 2 N-units). But by phases I and II of the parameter design, the SD of R has been substantially reduced, therefore the target value of R can be set to be less than 6 N-units.

For instance, if the target value of R is 5 N-units, then $J = \sqrt{5} * .06/3.33 = .30$, where 3.33 is the new SD of K. Hence $(R2 - R1) = 4 * (5 * .06/.30 + .30 * 3.33) = 8.00$. Here is a summary of different values of $(R2 - R1)$[8]:

Original value: 12.59
After phase I design: 10.83
After phase III design: 8.00

Note that this whole process of reducing variability in R is achieved *without* managerial intervention such as buying new machines, hiring and training a new work force, or things of that order. It's just mathematics.[9]

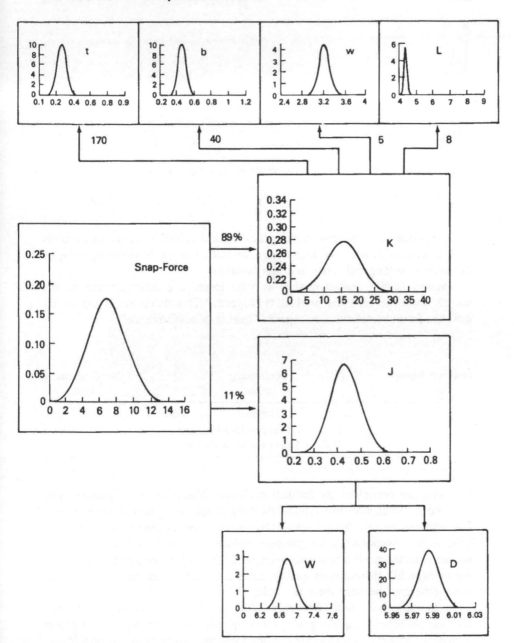

Figure 7 Hierarchy of the attribution of the variabilities in R, K, and J. (Reprinted from Oh and Im, *Robust Design, A Need and an Opportunity*, 1991, presented in the 218th Meeting of the Institute of Mathematical Statistics. Courtesy of Oh and Im.)

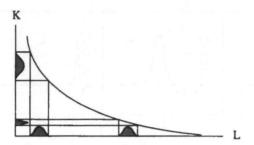

Figure 8   Exploitation of the inverse-square relationship in K vs. L.

•   •   •   •   •   •   •   •   •

In addition to parameter design, the Taguchi school suggests other tools such as system design, tolerance design, and other subskills including analysis of variance, orthogonal array, and loss function.

Taguchi (1981) defines quality as "the losses a product imparts to the society from the time the product is shipped." This definition is drastically different from the definition as regards "quality of conformance":

Nonconforming items		Conforming items		Nonconforming items
	LSL	Nominal	USL	

LSL = Lower specification limit
USL = Upper specification limit

This diagram resembles the football goalposts. Therefore some authors (see, e.g., Ross, 1988) label this type of thinking in QC as "goalpost philosophy." For instance, assume that the specifications for a special kind of metal bolt are 2.00 ± .02″. According to the goalpost philosophy, a bolt with diameter 2.02″ will be accepted while a bolt with diameter 2.021″ will be rejected. But according to Taguchi's definition of quality, there is probably little practical or functional difference between these two bolts.

In addition, bolts with diameters 2.02″ and 2.01″ are considered, by the goalpost philosophy, equally good in their quality. But in reality, the 2.01″ bolt is likely to impart less loss to the customer and the manufacturer as well.[10]

In the example of Oh and Im's fasteners, a goalposter might conclude that "there is nothing we can do about it," because variations in t,b,w,L,D,W have been reduced to the lowest levels current technology would allow. But as

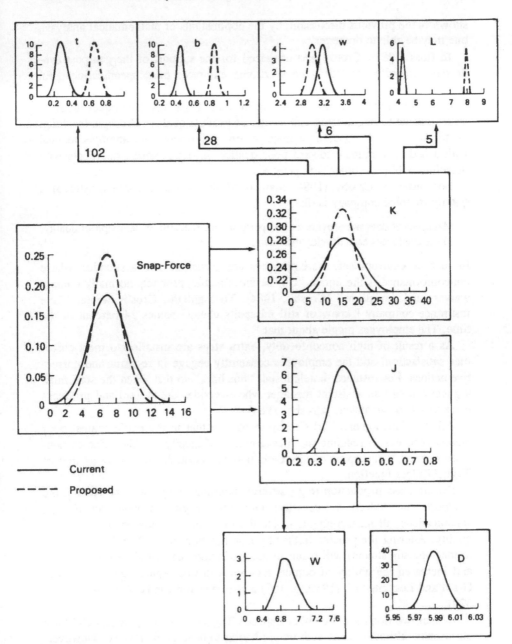

Figure 9 Reduction of variations in K and R. (Reprinted from Oh and Im, *Robust Design, A Need and an Opportunity*, 1991, presented in the 218th Meeting of the Institute of Mathematical Statistics. Courtesy of Oh and Im.)

shown in the previous discussion, by the applications of mathematical analysis, one may be able to do better.

In Ross (1988), Crosby was criticized for the support of the goalpost syndrome. According to Ross, the syndrome embraces print specifications, but ignores customer's requirements. In our opinion, this criticism is not well founded and indeed creates more confusion. Taguchi's techniques address issues related to engineering and design of product quality. Crosby's teaching, on the other hand, concerns a management philosophy that attempts to deal with a laid-back attitude toward poor quality. In other words, one deals with machines, while the other with people.

For instance, Crosby (1984) maintained that a major characteristic of a quality-troubled company is that:

> Management does not provide a clear performance standard or definition of quality, so the employees each develop their own.

In such an environment, the employees are accustomed to a situation where nonconformance is the norm: "Our services and/or products normally contain waivers and deviations" (Crosby, 1984). To highlight, Crosby wrote: "One insurance company I know of still misspells clients' names 24 percent of the time. The employees giggle about that."

As a result of high nonconformity, extra steps are installed to meet customer satisfaction, and the employees constantly engage in re-work and corrective actions. For instance, hotels install "hot lines" so that when the staff fails, a guest can call an assistant manager who overrides the system and produces extra towels or whatever. (Crosby, 1984).

All these phenomena lead Crosby to believe that in the battle against poor quality, one has to redefine the measurement of quality as "the price of nonconformance." The spirit of this definition is essentially the *same* as that of Taguchi's loss function.

During mass production (e.g., several thousand components per hour), the goalposts are indeed a necessity: Once the engineers have set up the specifications, all measurements inside the limits are considered equal in their quality. Altering the process (either by loss function or by "careful attention" to machine adjustment) will result in nonproductive labor, reduced production, and increased variability of output. Examples in this regard can be found in Grant and Leavenworth (1980, p. 107) and Evans and Lindsay (1989, p. 328–329).

In sum, we don't see a conflict between Taguchi's techniques and Crosby's philosophy, although some authors are trying to have us believe so. However, this conclusion is not an endorsement of the whole of Crosby's writings.[11] His teaching, for instance, may not provide much help to the Japanese or to companies that already strive for excellence and perfection.

Among all philosophies we have discussed so far, we haven't found any of them to be perfect. But the reader may query: Whose philosophy will produce the best result? Deming's or Crosby's? Or Ishikawa's? Our response to this question is statistical: You will probably never know for certain.

This answer may appear not to have said much, but it is certainly in accord with what Deming taught us in 1965: If a statistician was to make recommendations on administrative policy or decisions, he would cease to be a statistician.

NOTES

1. All projections were for bachelor's and master's degree students.
2. Deep in our hearts, all statisticians would hope the NSF projection is true. Twelve years have passed since the projection was made. There has been no follow-up study from NSF, and it seems evident that the projection was well off the mark.
3. Here is the exact quote: "Quality improvement means change. The rational basis for change is data. Data means statistical methods." This "paradigm," as Hahn and Boardman wrote, was from Bill Hunter.

   I talked to Hahn in a conference, and have read many of his writings. I personally hold high regard for his work and for his understanding of data, decision making, and statistics. Nevertheless, he made a statement that is misleading, therefore he deserves the criticism.
4. It has been proved (Diaconis and Freedman, 1988) that current nonparametric Bayesian models lead to inconsistent estimators. That means the greater the amount of data that are coming in, the more a Bayesian statistician will believe a wrong result.
5. If Deming's analysis is correct, other countries (or large corporations) should be able to replicate the "Japanese miracle." But so far most efforts in this direction have failed. My analysis boils down to two factors behind this "miracle": (1) the effective use of scientific methods, and (2) an almost "total commitment" of all parties involved in the battle against poor quality. This total commitment is in large part due to a unique Japanese culture that values education, a close link to self-esteem, and most importantly an unbelievable desire for perfection. This desire may be a result of the Samurai tradition that considered life to have only two meanings: It is *either* perfect *or* zero!
6. This 10%, albeit only a small portion, is crucial. As an analogy, consider a carton of orange juice, in which 90% (or more) is water. Without the remaining 10%, the orange juice would not be orange juice.
7. Deming was a physicist-turned-statistician. I doubt that most "applied statisticians" have a strong science background like Deming.
8. Theoretically, formulas (1–3) can be used an infinite number of times. But in reality, the process converges rapidly.
9. In some cases, the calculated range (R1,R2) may not fall within the tolerance region of R. If this is so, then other mathematical analysis will be needed.

Technically speaking, since K and J are both normally distributed, R = K ∗ J cannot be normal. But this is a relatively minor issue in this case study and will not be further discussed.

10. See also Taguchi (1981, pp. 10–14) for an insightful (but more technical) discussion of his loss function.

11. In his effort to demonstrate that the goal of zero-defects is attainable, Crosby (p. 75) wrote, "payroll doesn't make mistakes." The reason was that people wouldn't put up with it. Crosby's statement unfortunately is not true. For instance, in October 1988 the Central Payroll Office of the New Jersey State Government made a big mistake by giving every employee a double paycheck. The Office later corrected the error, but caused enormous problems for some of the employees.

REFERENCES

Anderson, T.W. and Sclove, S.L. (1986). *The Statistical Analysis of Data*, 2nd edition. The Scientific Press, South San Francisco.

Breiman, L. and Cutler, A. (1989). A Deterministic Algorithm for Global Optimization. Technical Report #224, University of California, Berkeley.

Crosby, P.B. (1984). *Quality without Tears: the Art of Hassle-Free Management*. McGraw-Hill, New York.

DeGroot, M.H. (1987). A Conversation with C.R. Rao. *Statistical Science*, Vol. 2, No. 1, 53–67.

Deming, W.E. (1960). *Sample Design in Business Research*. Wiley, New York.

Deming, W.E. (1938, 1943, 1964). *Statistical Adjustment of Data*. Dover, Mineola, New York.

Deming, W.E. (1965). Principles of Professional Statistical Practice. *Annals of Math. Stat.*, Vol. 36, 1883–1900.

Deming, W.E. (1978). Statistics Day in Japan. *American Statistician*, Vol. 32, No.4, p. 145.

Deming, W.E. (1982). *Quality, Productivity, and Competitive Position*. Massachusetts Institute of Technology, Center for Advanced Engineering Study.

Evans J.R. and Lindsay, W.M. (1989). *The Management and Control of Quality*. West Publishing Company, Saint Paul, MN.

Godfrey, A.B. (1985). Comment on "The Key Role of Statisticians in the Transformation of North American Industry" by B.L. Joiner. *American Statistician*, Vol. 39, No. 3, 231–232.

Grant, E.L. and Leavenworth, R.S. (1980). *Statistical Quality Control*, 5th edition. McGraw-Hill, New York.

Hahn and Boardman (1987). The Statistician's Role in Quality Improvement. *AMSTAT News*, March, 5–8.

Hoaglin, D.C., Mosteller, F., and Tukey, J.W. (1985, ed.). *Exploring Data Tables, Trends, and Shapes*. Wiley, New York.

Ishikawa, K. (1985). *What is Total Quality Control? The Japanese Way*. Prentice Hall, Englewood Cliffs, New Jersey.

Joiner, B.L. (1985). The Key Role of Statisticians in the Transformation of North American Industry. *American Statistician*, Vol. 39, no. 3, 224–234.

Juran, J.M. (1951/1979). *Quality Control Handbook*, 3rd edition. McGraw-Hill, New York.

Marquardt, D.W. (1987). The Importance of Statisticians. *JASA*, Vol. 82, No. 397, 1–7.

Oh, H.L. and Im, K.Y. (1991). Robust Design, a Need and An Opportunity. Paper presented to 218th Meeting of the Institute of Mathematical Statistics, Philadelphia.

Ross, P.J. (1988). *Taguchi Techniques for Quality Engineering*. McGraw-Hill, New York.

Shafer, G. (1987). Probability Judgment in Artificial Intelligence and Expert Systems. *Statistical Science*, Vol. 2, No. 1, 3–16.

Spiegelhalter, D.J. (1987). Probabilistic Expert Systems in Medicine: Practical Issues in Handling Uncertainty. *Statistical Science*, Vol. 2, No. 1, 25–30.

Taguchi, G.(1981). *On-line Quality Control during Production*. Published by Japanese Standards Association.

Tamura, M. (1989). Statistical Models and Analysis in Auditing. *Statistical Science*, Vol. 4, No. 1, 2–33.

Tukey, J.W. (1977). *Exploratory Data Analysis*. Addison-Wesley, Reading, Massachusetts.

Epilog: Toward a New Perspective on Statistical Inference

Sometimes one may wonder: "What is statistics really about?" If you pose this question to researchers from different fields, very likely you will get different answers. A quantitative behavioral scientist may tell you that without a statistical test you won't be able to get your papers published. On the other hand, an engineer may say, "Statistics? There is nothing in it!"[1]

Among statisticians, it is widely held that statistics is a science that relates to experiments and the quantification of uncertainty and thus provides guiding light for research workers and decision-makers. This is true. However, Bradley (1982, *JASA*) observed a disheartening fact. In a centennial issue of *Science* (1980), where scholarly discussions ranged from "Present and Future Frontiers of the Science," to "The Interaction of Science and Technology with Societal Problems," nowhere did *Science* magazine mention statistics as a discipline of science.

In brief, statistics is a discipline that is neglected by certain hard scientists but at the same time overused (and misused) by quantitative researchers in soft sciences such as economics, earth science, psychology, education, and social science. No wonder the discipline is perceived by some as a backwater of scholarly activity. To restore self-respect to this troubled discipline, the statistical community has a long way to go. It must, for example, stamp out the misuse of statistics in soft sciences (and in statistical literature as well). After all, a big portion of misused statistics happen simply because statisticians allow them to happen.

233

In addition, we should tell the statistics users that essentially there are only four components in statistics that are deductive in their nature:

1. randomized experiments,
2. randomized sample surveys,
3. mathematical statistics and probability, and
4. regression analysis where the functional form of the model is dictated by strong theory or by a controlled experiment.

Anything else in statistical inference is inductive and can be highly speculative.

Inductive (i.e., speculative) methods of statistical inference consist of a large variety of techniques. Here is a short list:

1. regression analysis in non-experimental sciences,
2. P-values and confidence intervals for non-randomized data sets,
3. time-series models, and
4. Bayesian analysis.

Such methods sometimes *rival* those in randomized studies, because often, "Nature has done the randomization for us" (Fisher, 1974).

But more often, such methods are the source of corrupt statistics, because, "Nature was not in the habit of doing the randomization" (Hinkeley, 1980). In fact, much statistics in this area share a common characteristic: they look like math but are not math; they look like science but are not science.

To avoid self-deception, Diaconis (1985) proposed seven remedies that can be used to deal with problems posed by such analyses. The first remedy is "Publish without P-values," and if confirmatory and speculative analyses appear in the same document, "the two efforts should be clearly distinguished."[2]

There is another group of statistical techniques that are known as descriptive statistics and EDA (exploratory data analysis). Such techniques do not assume probability models in the inference. But in numerous cases, as noted in Diaconis (1985), EDA has proven more effective than classical techniques that rely on probabilistic assumptions.[3] Again some notes of caution: (1) the nature of EDA is speculative; and (2) its interpretation and the related decision-making are quite personal and subjective.

It is important to note that there is nothing wrong with subjective knowledge, as long as one knows what one is doing. In the fields of non-experimental (or semi-experimental) sciences, most knowledge is speculative in its nature. This does not mean subjective knowledge is not useful. On the contrary, for most decision-makers, knowledge from social science and psychology (and even from philosophy) will be more useful than that from hard sciences (e.g., quantum mechanics).

Soft scientists (such as psychologists, educators, or social scientists) often complain that they are looked down upon by mathematicians and mathematical statisticians. This is true, but the main reason is *not* that soft scientists are soft in mathematics. Rather, it is because soft scientists often use statistical methods in a religious fashion, and drape their subjective knowledge under the cover of "objective science." In this type of "science," research workers have been taught (see examples in Chapter 4) to follow procedures that are "totally objective" but indeed utterly mindless.

A reason that statistics users are so ignorant is that the nature of "data analysis" is seldom made clear in popular textbooks—as in old-fashioned capitalism, they leave the responsibility to the consumers. For such reasons, we unabashedly volunteer to provide a new perspective toward statistical inference in science and decision-making. To begin with, let's consider the Venn diagram (Fig. 1).

For statistics users, the good news is that the same statistical formulas work for A, B, and C in Fig. 1; but the bad news is that these formulas do not tell a user anything about whether he is located in A, B, or C.

In order to be in category A, a statistical analysis has to be preceded by randomization or a controlled experiment. In soft sciences, such justifications are not possible. Then one has to assess

1. the randomness of the data,
2. the stability of the phenomenon under study, and
3. the homogeneity of the population.

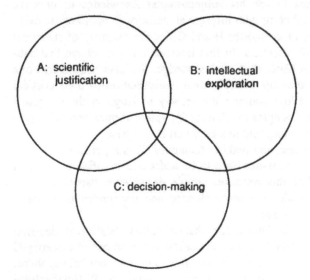

Figure 1   Venn diagram for different aspects of statistical inference.

The majority of "statistical justifications" in soft sciences (such as t-test or ANOVA) do not meet these requirements and have to be looked at with reservations, or, in our opinion, be thrown out of category A altogether.

Nevertheless, research findings in soft sciences can be valuable for intellectual exploration (category B) and also for decision making (category C). In these categories, statistical inference has been used extensively; and it is *not* a statistician's business to condemn such use. On the contrary, promotion of statistical methods in these areas would help the statistical community and our society as a whole. Further, if statisticians are willing to take the soft sciences to heart, then the borrowed strength from other disciplines would enrich the teaching and the practice of statistical methods.

Category B is an intellectual territory that has no limit. In this territory, an investigator should allow his imagination to run wild without the fear that his beautiful theory may one day be ridiculed by an ugly experiment. In this regard, mathematical statistics and probability (especially stochastic processes) can be powerful vehicles for theory development, as shown in the frontiers of hard sciences such as quantum mechanics, astrophysics, engineering, biomedical science, and in the new field of chaos, dynamical systems, and wavelet transform. Such mathematical constructs are infinitely rich in their subtleties and esoteric beauty, a trademark of the statistical profession.[4]

By contrast, in soft sciences and in decision-making, formal statistical inferences are surprisingly irrelevant to most data one usually encounters. The reason is that in these disciplines, measurements are primitive (or necessarily primitive), hence "facts" are tentative, fragile, and easily crushed. In such cases, the investigator has to use his subject-matter knowledge to improve upon his measurements, dabbling with inferential statistics is not likely to help.

Within the intersection of categories B and C lies the majority of statistical analyses generated by soft scientists. In this intersection, we believe that subjective knowledge is often better than no knowledge. We also believe that one should not be afraid to come up with his own interpretation of the statistical artifacts. In fact, if you allow yourself the leeway to laugh at these "quasi-inferential statistics" (see examples in Chapter 2), you also open yourself up to the possibilities when interesting findings take you by surprise.

Note that in most cases decision-making (category C) is a personal and subjective activity. In such cases, issues have to be addressed in different contexts and in different ways. For this purpose, simple descriptive statistics are of great value. But such statistics are best collected and interpreted by subject-matter experts, not by statisticians.

As a sidenote, a body of knowledge that is called "statistical decision theory" (see, e.g., Berger, 1980) is often classified as a member of category C (decision-making). This is a mistake. The thing simply does not belong there. Rather, it belongs to the upper-right corner of category B (intellectural exploration).

It should be pointed out that statisticians do contribute *immensely* to decision-makers. But this contribution is accomplished (1) by providing *reliable* information (the intersection of A and C) through randomized experiments or sample surveys, (2) by weeding out shoddy statistics that, if taken at their face values, would lead to disaster, and (3) by providing quasi-inferential statistics to decision-makers when better alternatives do not exist.

• • • • • • • • •

In non-experimental sciences, adding numbers allows easy judgment. But in order to earn nods from hard-line statisticians, a Bayesian spirit is needed. Here by *Bayesian spirit* we mean a relentless pursuit of prior knowledge other than that revealed by the data in hand. This intellectual pursuit includes brainstorming, scholarly exchanges, and non-emotional debates.

Some scholars may feel uncomfortable with sharp language in heated debates that may touch the soft spots of personality. But it is precisely this uncomfortable state in which oysters produce *pearls*.

Another reservation regarding debates is that debates may be full of anecdotal evidence and personal opinions. But so what? Anecdotal evidence appears foolish only to fools.[5]

By all accounts, academicians are among the most reasonable human beings. Given a host of information, they usually know what is scientific (or at least what is not), *provided* that a thorough debate has taken place.

We believe that if there were more scholarly exchanges among academicians, there would not exist so much shoddy statistics masquerading as scientific evidence.

Finally, we would like to complement a statement put together by Dr. Barbara Bailar, executive director of the American Statistical Association,

> Statistics affect all aspects of our lives. Government economic statistics affect wages for workers; medical care is affected by health statistics. There really isn't a part of our lives that's not affected by statistics. (*The New York Times*, August 7, 1989)

This is certainly true. But in keeping up with the theme of this book, let's remind statistics users that our society and our lives are affected by *both* correct and corrupt statistics.

In addition, it is not only our responsibility to point out the misuse of statistics, but also it is a *thrill* when you (or your students) uncover a professional slip committed by a famous scientist (or statistician). The true spirit of science is self-correcting, and there can never be too much of it.

NOTES

1. This is exactly what I have heard from some engineers.
2. For instance, in a two-sample comparison, one can report means, standard deviations, and sample sizes, but not standard errors or P-values.

3. If one does not rule out probability models from EDA (as Tukey originally intended), then one may also put regression and time-series models in a subcategory of exploratory data analysis.

4. Some empirical scientists are hostile to theorists. But the history of science has demonstrated this point over and over again: In order to make progress, any scientific discipline needs people who conduct "desk research" on seemingly useless problems. A prominent example in this category is the breakthrough via Dirac's equation in quantum-relativity theory (see the Introduction to Chap. 6). Another example is the Cantor set that had long been perceived as "pathological" in measure theory but is now widely used by chaoticians in physics and other scientific disciplines.

5. This is a paraphrase of a statement in Deming (1982, p. 99): "Wisdom sounds foolish to fools."

REFERENCES

Berger, J.O. (1980). *Statistical Decision Theory: Foundations, Concepts, and Methods*, Springer-Verlag, New York.

Bradley, R. (1982). The Future of Statistics as a Discipline (Presidential Address). *JASA*, Vol. 77, No. 377, 1–10.

Deming, W.E. (1982). *Quality, Productivity, and Competitive Position*. Massachusetts Institute of Technology, Center for Advanced Engineering Study.

Diaconis, P. (1985). Theories of Data analysis: From Magical Thinking Through Classical Statistics, in *Exploring Data Tables, Trends, and Shapes*, edited by D.C. Hoaglin, F. Mosterller, and J.W. Tukey. Wiley, New York.

Fisher, R.A. (1974). *Collected Papers of R.A. Fisher*. Adelaide, Australia, University of Adelaide.

Hinkeley, D.V. (1980). Comment to "Randomization Analysis of Experimental Data: The Fisher Randomization Test" by D. Basu. *JASA*, Vol. 75, No. 371, 582–584.

Index